Edwin J. (Edwin James) Houston, Arthur E. (Arthur Edwin) Kennelly

Electro-Dynamic Machinery for Continuous Currents

Edwin J. (Edwin James) Houston, Arthur E. (Arthur Edwin) Kennelly
Electro-Dynamic Machinery for Continuous Currents
ISBN/EAN: 9783337163716

Printed in Europe, USA, Canada, Australia, Japan

Cover: Foto ©berggeist007 / pixelio.de

More available books at **www.hansebooks.com**

ELECTRO-DYNAMIC MACHINERY

FOR CONTINUOUS CURRENTS

BY

EDWIN J. HOUSTON, Ph. D. (Princeton)

AND

A. E. KENNELLY, Sc. D.

NEW YORK
THE W. J. JOHNSTON COMPANY
253 Broadway
1896

PREFACE.

ALTHOUGH several excellent treatises on machinery employed in electro-dynamics already exist, yet the authors believe that there remains a demand for a work on electro-dynamic machinery based upon a treatment differing essentially from any that has perhaps yet appeared. Nearly all preceding treatises are essentially symbolic in their mathematical treatment of the quantities which are involved, even although such treatment is associated with much practical information. It has been the object of the authors in this work to employ only the simplest mathematical treatment, and to base this treatment, as far as possible, on actual observations, taken from practice, and illustrated by arithmetical examples. By thus bringing the reader into intimate association with the nature of the quantities involved, it is believed that a more thorough appreciation and grasp of the subject can be obtained than would be practicable where a symbolic treatment from a purely algebraic point of view is employed.

In accordance with these principles, the authors have inserted, wherever practicable, arithmetical examples, illustrating formulas as they arise.

The fundamental principles involved in the construction and use of dynamos and motors have been considered, rather than the details of construction and winding.

The notation adopted throughout the book is that recommended by the Committee on Notation of the Chamber of Delegates at the Chicago International Electric Congress of 1893.

The magnetic units of the C. G. S. system, as provisionally adopted by the American Institute of Electrical Engineers, are employed throughout the book.

The advantages which are believed to accrue to the conception of a working analogy between the magnetic and voltaic circuits, are especially developed, for which purpose the conception of reluctivity and reluctance are fully availed of.

CONTENTS.

CHAPTER I.

GENERAL PRINCIPLES OF DYNAMOS.

Definition of Electro-Dynamic Machinery. General Laws of the Generation of E. M. F. in Dynamos. Electric Capability. Output. Intake. Commercial Efficiency. Electrical Efficiency. Maximum Output. Maximum Efficiency. Relation between Output and Efficiency, . 1

CHAPTER II.

STRUCTURAL ELEMENTS OF DYNAMO-ELECTRIC MACHINES.

Armatures. Field Magnets. Magnetic Flux. Commutator Brushes. Constant-Potential Machines. Constant-Current Machines. Magneto-Electric Machines. Separately-Excited Machines. Self-Excited Machines. Series-Wound Machines. Shunt-Wound Machines. Compound-Wound Machines. Bipolar Machines. Multipolar Machines. Quadripolar, Sextipolar, Octopolar and Decipolar Machines. Number of Poles Required for Continuous and Alternating-Current Machines. Consequent Poles. Ring Armatures. Drum Armatures. Disc Armatures. Pole Armatures. Smooth-Core Armatures. Toothed-Core Armatures. Inductor Dynamos. Diphasers. Triphasers. Single Field-Coil Multipolar Machines. Commutatorless Continuous-Current Machines, ... 9

CHAPTER III.

MAGNETIC FLUX.

Working Theory Outlined. Magnetic Fields. Direction, Intensity, Distribution. Uniformity, Convergence, Divergence. Flux Density. Tubes of Force. Lines of Magnetic Force. The Gauss. Properties of Magnetic Flux. M. M. F. Ampere-Turn. The Gilbert. Flux Paths, ... 29

CHAPTER IV.

NON-FERRIC MAGNETIC CIRCUITS.

Reluctance. The Oersted. Ohm's Law Applied to Magnetic Circuits. Ferric, Non-Ferric, and Aero-Ferric Circuits. Magnetizing Force. Magnetic Potential. Laws of Non-Ferric Circuits, ... 48

CHAPTER V.

FERRIC MAGNETIC CIRCUIT.

Residual Magnetism. Permeability. Theory of Magnetization in Iron. Prime M. M. F. Structural M. M. F. Counter M. M. F. Reluctivity. Laws of Reluctivity. 55

CHAPTER VI.
AERO-FERRIC MAGNETIC CIRCUITS.
Magnetic Stresses.   Laws of Magnetic Attraction.   Leakage, . . . 68

CHAPTER VII.
LAWS OF ELECTRO-DYNAMIC INDUCTION.
Fleming's Hand Rule. Cutting and Enclosure of Magnetic Flux, . . 74

CHAPTER VIII.
ELECTRO-DYNAMIC INDUCTION IN DYNAMO ARMATURES.
Curves of E. M. F. Generated in Armature Windings. Idle-Wire, . 90

CHAPTER IX.
ELECTROMOTIVE FORCE INDUCED BY MAGNETO GENERATORS, 103

CHAPTER X.
POLE ARMATURES, 110

CHAPTER XI.
GRAMME-RING ARMATURES.
E. M. Fs. Induced in. Effect of Magnetic Dissymmetry. Commutator-Brushes. Effect of Dissymmetry in Winding. Best Cross-Section of Armature, ... 117

CHAPTER XII.
CALCULATION OF THE WINDINGS OF A GRAMME-RING DYNAMO, 128

CHAPTER XIII.
MULTIPOLAR GRAMME-RING DYNAMOS.
Belt-Driven *versus* Direct-Driven Generators. Reasons for Employing Multipolar Field Magnets. Multipolar Armature Connections. Effect of Dissymmetry in Magnetic Circuits of Multipolar Generators. Computations for Multipolar Gramme-Ring Generator, ... 135

CHAPTER XIV.
DRUM ARMATURES.
Smooth-Core and Toothed-Core Armatures. Armature Windings. Lap Windings. Wave Windings, ... 152

CHAPTER XV.
ARMATURE JOURNAL BEARINGS.
Frictional Losses of Energy in Dynamos. Sight-Feed Oilers and Self-Oiling Bearings, ... 159

CHAPTER XVI.

EDDY CURRENTS.

Methods of Lamination of Core. Transposition of Conductors, . . 164

CHAPTER XVII.

MAGNETIC HYSTERESIS.

Nature and Laws of Hysteresis. Hysteretic Loss of Energy. Table of Hysteretic Loss. Hysteretic Torque, ... 172

CHAPTER XVIII.

ARMATURE REACTION AND SPARKING AT COMMUTATORS.

Diameter of Commutation. E. M. F. of Self-Induction. Inductance of Coils. Cross-Magnetization. Back-Magnetization. Leading and Following Polar Edges. Lead of Brushes. Distortion of Field. Conditions Favoring Sparking at Commutator. Conditions Favoring Sparkless Commutation. Methods Adopted for Preventing Sparking, 179

CHAPTER XIX.

HEATING OF DYNAMOS.

Losses of Energy in Magnetizing, Eddies, Hysteresis and Friction. Safe Temperature of Armatures, ... 199

CHAPTER XX.

REGULATION OF DYNAMOS.

Series-Wound, Shunt-Wound and Compound-Wound Generators. Overcompounding. Characteristic Curves of Machines. Internal and External Characteristic. Computation of Characteristics. Field Rheostats. Series-Wound Machines and their Regulation. Open-Coil and Closed-Coil Armatures, ... 206

CHAPTER XXI.

COMBINATIONS OF DYNAMOS IN SERIES AND PARALLEL.

Generator Units. Series-Wound Machines Coupled in Series. Shunt-Wound Machines Coupled in Parallel. Equalizing Bars. Omnibus Bars, ... 220

CHAPTER XXII.

DISC-ARMATURES AND SINGLE-FIELD COIL MACHINES, 228

CHAPTER XXIII.

COMMUTATORLESS CONTINUOUS-CURRENT GENERATORS.

Disc and Cylinder Machines, ... 234

CONTENTS.

CHAPTER XXIV.
ELECTRO-DYNAMIC FORCE.
Fleming's Hand-Rule. Ideal Electro-dynamic Motor, ... 241

CHAPTER XXV.
MOTOR TORQUE.
Torque of Single Active Turn. Torque of Armature-Windings. Torque of Multipolar Armatures. Dynamo-Power, ... 251

CHAPTER XXVI.
EFFICIENCY OF MOTORS.
Commercial Efficiency in Generators and Motors Compared. Slow-Speed *versus* High-Speed Motors. Torque-per-pound of Weight, . . 268

CHAPTER XXVII.
REGULATION OF MOTORS.
Control of Speed and Torque under Various Conditions. Control of Series-Wound Motors, ... 280

CHAPTER XXVIII.
STARTING AND REVERSING OF MOTORS.
Starting Rheostats. Starting Coils. Automatic Switches. Direction of Rotation in Motors, ... 297

CHAPTER XXIX.
METER-MOTORS.
Conditions under which Motors may act as Meters, ... 309

CHAPTER XXX.
MOTOR DYNAMOS.
Construction and Operation of Motor-Dynamos, ... 318

ELECTRO-DYNAMIC MACHINERY

FOR CONTINUOUS CURRENTS.

CHAPTER I.

GENERAL PRINCIPLES OF DYNAMOS.

1. By electro-dynamic machinery is meant any apparatus designed for the production, transference, utilization or measurement of energy through the medium of electricity. Electro-dynamic machinery may, therefore, be classified under the following heads :

(1.) Generators, or apparatus for converting mechanical energy into electrical energy.

(2.) Transmission circuits, or apparatus designed to receive, modify and transfer the electric energy from the generators to the receptive devices.

(3.) Devices for the reception and conversion of electric energy into some other desired form of energy.

(4.) Devices for the measurement of electric energy.

Under generating apparatus are included all forms of continuous or alternating-current dynamos.

Under transmission circuits are included not only conducting lines or circuits in their various forms, but also the means whereby the electric pressure may be varied in transit between the generating and the receptive devices. This would, therefore, include not only the circuit conductors proper, but also various types of transformers, either stationary or rotary.

Under receptive devices are included any devices for converting electrical energy into mechanical energy. Strictly speaking, however, it is but fair to give to the term mechanical energy a wide interpretation, such for example, as would permit the introduction of any device for translating electric energy into telephonic or telegraphic vibrations.

Under devices for the measurement of electric energy would be included all electric measuring and testing apparatus.

In this volume the principles underlying the construction and use of the apparatus employed with continuous-current machinery will be considered, rather than the technique involved in their application.

2. A consideration of the foregoing classification will show that in all cases of the application of electro-dynamic machinery, mechanical energy is transformed, by various devices, into electric energy, and utilized by various electro-receptive devices connected with the generators by means of conducting lines. The electro-technical problem, involved in the practical application of electro-dynamic machinery, is, therefore, that of economically generating a current and transferring it to the point of utilization with as little loss in transit as possible. The engineering problem is the solution of the electro-technical problem with the least expense.

3. A *dynamo-electric generator* is a machine in which conductors are caused to cut *magnetic flux-paths*, under conditions in which an expenditure of energy is required to maintain the electric current. Under these conditions, electromotive forces are generated in the conductors.

Since the object of the electromotive force generated in the armature is the production of a current, it is evident that, in order to obtain a powerful current strength, either the electromotive force of the generator must be great, or the resistance of the circuit small.

Electromotive sources must be regarded as primarily producing, not electric currents, but electromotive forces. Other things being equal, that type of dynamo will be the best electrically, which produces, under given conditions of resistance, speed, etc., the highest electromotive force (generally contracted E. M. F.). In designing a dynamo, therefore, the electromotive force of which is fixed by the character of the work it is required to perform, the problem resolves itself into obtaining a machine which will satisfactorily perform its work at a given efficiency, and without overheating, with, however, the maximum economy of construction and operation. In other words, that dynamo will be the best, electrically, which for a given weight, resistance and friction, produces the greatest electromotive force.

4. There are various ways in which the electromotive force of a dynamo may be increased; viz.,

(1.) By increasing the speed of revolution.
(2.) By increasing the magnetic flux through the machine.
(3.) By increasing the number of turns on the armature.

The increase in the speed of revolution is limited by well-known mechanical considerations. Such increase in speed means that the same wire is brought through the same magnetic flux more rapidly. To double the electromotive force from this cause, we require to double the rate of rotation, which would, in ordinary cases, carry the speed far beyond the limits of safe commercial practice.

Since the E. M. F. produced in any wire is proportional to its rate of cutting magnetic flux, it is evident that in order to double the E. M. F..in a given wire or conductor, its rate of motion through the flux must be doubled. This can be done, either by doubling the rapidity of rotation of the armature; or, by doubling the density of the flux through which it cuts, the rate of motion of the armature remaining the same.

Since the total E. M. F. in any circuit is the sum of the separate E. M. Fs. contained in that circuit, if a number of separate wires, each of which is the seat of an E. M. F., be connected in series, the total E. M. F. will be the sum of the separate E. M. Fs. If, therefore, several loops of wire be moved through a magnetic field, and these loops be connected in series, it is evident that, with the same rotational speed and flux density, the E. M. F. generated will be proportional to the number of turns.

An increase in E. M. F. under any of these heads is limited by the conditions which arise in actual practice. As we have already seen, the speed is limited by mechanical considerations. An increase in the magnetic flux is limited by the *magnetic permeability* of the iron—that is, its capability of conducting magnetic flux—and the increase in the number of turns is limited by the space on the armature which can properly be devoted to the winding.

5. It will be shown subsequently that a definite relation exists between the output of a dynamo, and the relative amounts of iron and copper it contains—that is to say, the type of machine being determined upon, given dimensions and weight should produce, at a given speed, a certain output. The conditions under which these relations exist will form the subject of future consideration.

6. Generally speaking, in the case of every machine, there exists a constant relation between its electromotive force and resistance, which may be expressed by the ratio, $\frac{E^2}{r}$, where E, is the E. M. F. of the machine at its brushes, in volts, and r, the resistance of the machine; *i. e.*, its internal resistance, in ohms. In any given machine, the above ratio is nearly constant, no matter what the winding of the machine may be; *i. e.*, no matter what the size of the wire employed.* This ratio may be taken as representing, in watts, the electric activity of the machine on short circuit, and may be conveniently designated the *electric capability* of the machine. For example, in a 200, KW (200,000 watts) machine; *i. e.*, a dynamo, whose output is 200 KW (about 267 horse power), the value of the electric capability would be about 10,000 KW, so that, since $\frac{E^2}{r} = 10,000,000$, if its E. M. F. were 155 volts, its resistance would be 0.0024 ohm; whereas, if its E. M. F. were 100 volts, its resistance would be approximately 0.001 ohm.

7. Hitherto we have considered the energy absorbed by the dynamo, independently of its external circuit—that is, we have considered only the electric capability of the machine.

When the dynamo is connected with an external circuit, two extreme cases may arise; viz.,

---

* This ratio would be constant if the ratio of insulation thickness to diameter of wire remained constant through all sizes of wire.

(1.) When the resistance of the external circuit is very small, so that the machine is practically short circuited. Here all the electric energy is liberated within the machine.

(2.) When the external resistance is so high that the resistance of the machine is negligible in comparison. Here practically all the energy in the circuit appears outside the machine. The total amount of work, however, performed by the machine, under these circumstances, would be indefinitely small, since the current strength would be indefinitely small. Between these two extreme cases, an infinite number of intermediate cases may arise.

8. By the *output* of a dynamo is meant the electric activity of the machine in watts, as measured at its terminals; or, in other words, the output is all the available electric energy. Thus, if the dynamo yields a steady current strength of 500 amperes at a steady pressure or E. M. F., measured at its terminals, of 110 volts, its output will be $110 \times 500 = 55,000$ watts, or 55 kilowatts.

The *intake* of a dynamo is the mechanical activity it absorbs, measured in watts. Thus, if the dynamo last considered were driven by a belt, which ran at a speed of 1,500 feet-per-minute, or 25 feet-per-second, and the tight side of the belt exerted a stress or pull of 2,500 pounds weight, while the slack side exerted a pull of 710 pounds weight, the effective force, or that exerted in driving the machine, would be 1,790 pounds weight. This force, moving through a distance of 25 feet per second, would develop an activity represented by $1,790 \times 25 = 44,750$ foot-pounds per second; and one foot-pound per second is usually taken as 1.355 watts, so that the intake of the machine is 60,630 watts, or 60.63 KW.

By the *commercial efficiency* of a dynamo is meant the ratio of its output to its intake. In the case just considered, the commercial efficiency of the machine would be $\frac{55}{60.63} = 0.9072$.

By the *electric efficiency* of a dynamo is meant the output, divided by the total electric activity in the armature circuit. Thus, if the dynamo just considered had a total electric energy in its circuit of 57 KW, of which 2 KW were expended in the machine, its electric efficiency would be $\frac{55}{57} = 0.965$.

9. The output of a machine would be greatest when the external resistance is equal to the resistance of the machine. In this case, the output would be just one-quarter the electric capability, and the electric efficiency would be 0.5. Thus, the resistance of the dynamo considered in the preceding paragraph would be, say, 0.008 ohm, and the electric capability of the machine $\frac{110^2}{0.008} = 1,512,500$ watts, or 1,512.5 KW. If the external resistance were equal to the internal resistance—namely, 0.008 ohm, the total activity in the circuit would be 756.25 KW; the output would be 378.12 KW, and the electric efficiency 0.5.

That is to say, in order to obtain a maximum output from a dynamo machine, the circumstances must be such that half the electric energy is developed in the machine, and half in the external circuit; or, in other words, the electric efficiency can be only 0.5. In practice, however, it would be impossible to operate a machine of any size under these circumstances, since the amount of energy dissipated in the machine would be so great that the consequent heating effects might destroy it.

10. We have seen that whenever the resistance in the external circuit is indefinitely great, as compared with that of the machine, the electric efficiency of the machine will be 1.0 or 100 per cent. It is evident, therefore, that in order to increase the electric efficiency of a dynamo, it is necessary that the resistance of the external circuit be made great, compared with the internal resistance of the machine. For example, if the external resistance be made nine times greater than that of the internal circuit, then the electric efficiency will be $\frac{9}{9+1} = 0.9$; and, similarly, if the external resistance be nineteen times that of the internal resistance, the electric efficiency would be raised to $\frac{19}{19+1} = 0.95$. Generally speaking, therefore, a high electric efficiency requires that the internal resistance of the machine be small as compared with the external resistance, and, also, that the amount of power expended in local circuits, as in magnetizing the field magnets of the dynamo, be relatively small.

11. Care must be taken not to confound the electric efficiency of a machine with its electric output. The electric output of a machine would reach a maximum when the electric efficiency was 0.5 or 50 per cent., and the output would be zero when the electric efficiency reached 1.0.

The electric efficiency of the largest dynamos is very high, about 0.985. Indeed, the electric efficiency of large machines must necessarily be made high, since, otherwise, the liberation of energy within them would result in dangerous overheating.

The commercial efficiency of a dynamo is always less than its electric efficiency, since all mechanical and magnetic frictions, such as air resistance, journal-bearing friction, hysteresis and eddy currents come into account among the losses. The commercial efficiency also depends upon the type of machine, whether it be belt-driven, or directly mounted on the engine shaft, since the mechanical frictions to be overcome differ markedly in these two cases. The commercial efficiency will also vary with the character of the iron employed in its field magnets and armature, and with the care exercised in securing its proper lamination. In large machines, of say 500 KW capacity, the commercial efficiency may be as high as 0.95. In very small machines, of say 0.5 KW, the highest commercial efficiency may be only 0.6.

12. Although in the United States it is the practice among constructors generally, to calculate, express and compare lengths and surface areas in inches and square inches, when referring to dynamo machinery, and although it might seem therefore more suitable to adopt inches and square inches as units of length and surface throughout this book; yet the fact that the entire international system of electro-magnetic measurement is based on the centimetre, renders the centimetre and square centimetre the natural units of dimensions in electro-magnetism. The authors have therefore preferred to base the formulæ and reasoning in this volume on the French fundamental units, in order to simplify the treatment, well knowing that once the elementary principles have been grasped, the transition to English measurements is easily effected by the student. The following data will, therefore, be useful:

1 inch = 2.54 cms.	1 cm. = 0.3937 inch.
1 foot = 30.48 cms.	1 cm. = 0.03281 foot.
1 sq. inch = 6.4515 sq. cms.	1 sq. cm. = 0.155 sq. in.
1 cubic inch = 16.387 c. c.	1 c. c. = 0.06102 c. in.

CHAPTER II.

STRUCTURAL ELEMENTS OF DYNAMO-ELECTRIC MACHINES.

13. Dynamo-electric machines, as ordinarily constructed, consist essentially of the following parts; namely,

(1.) Of the part called the *armature*, in which the E. M. F. is generated. The armature is generally a rotating part, although in some machines the armature is fixed, and either the field magnets, or the magnetic field, revolve.

(2.) Of the part in which the magnetic field is generated. This part is called the *field magnet* and provides a *magnetic flux* through which the conductors of the armature are generally, actually, and always relatively, revolved.

(3.) Of the part or parts that are employed for the purpose of collecting and rectifying the currents produced by the E. M. F. generated in the armature; *i. e.*, collecting and commuting them, and causing them to flow in one and the same direction in the external circuit. This portion is called the *commutator*.

(4.) Bundles of wire, metallic plates, metallic gauze, or plates of carbon, pressed against the commutator, and connected with the circuit in which the energy of the machine is utilized. These are called the *brushes*.

In addition to the above parts, which are directly connected with the electric actions of the machine, there are the necessary mechanical parts, such as the bearings, shaft, keys, base, etc., which also require attention.

The particular arrangement of the different parts will necessarily depend upon the type of machine, as well as on the character of the circuit which the machine is designed to supply.

It will, therefore, be advisable to arrange dynamo-electric machines into general classes, before attempting to describe the structure and peculiarities of their various parts.

14. Dynamos may be conveniently divided into the following classes; viz.,

(1.) *Constant-potential* machines, or those designed to maintain at their terminals a practically uniform E. M. F. under all variations of load.

To this class belong nearly all dynamos for supplying incandescent lamps and electric railroads.

Fig. 1 represents a particular machine of the constant-potential type. *A, A,* is the armature, whose shaft revolves in the self-oiling bearings *B, B. C* is the commutator, and *D, D,* are triple sets of brushes pressing their tips or ends upon

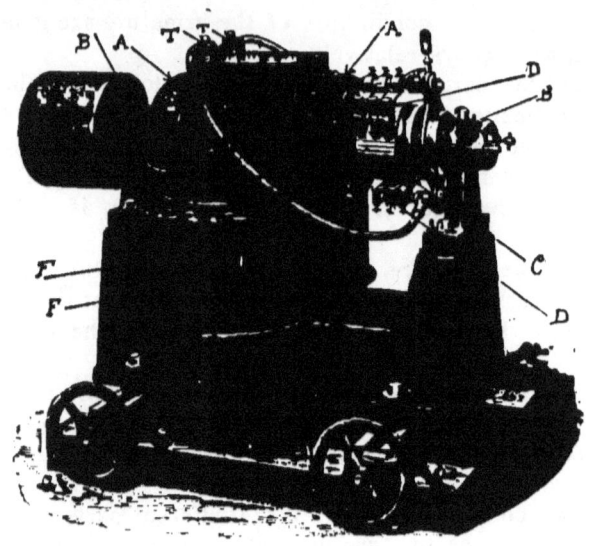

FIG. 1.—CONTINUOUS-CURRENT BIPOLAR CONSTANT-POTENTIAL GENERATOR.

the commutator. *F, F*, are the field magnets, wound with coils of insulated wire. *T, T*, are the machine terminals, connected with the brushes and with the external circuit or load. The whole machine rests on slides with screw adjustment for tightening the driving belt.

Constant-potential generators are made of all sizes, and of various types.

(2.) *Constant-current* machines, or those designed to maintain an approximately constant current under all variations of load.

Constant-current machines are employed almost exclusively for supplying arc lamps in series.

Fig. 2 represents a form of constant-current generator. This is an arc-light machine. It has four field magnets but only two poles, P^1 and P^2, connected by a bridge of cast iron at B. At R, is a regulating apparatus for automatically maintaining the constancy of the current strength, by rotating the

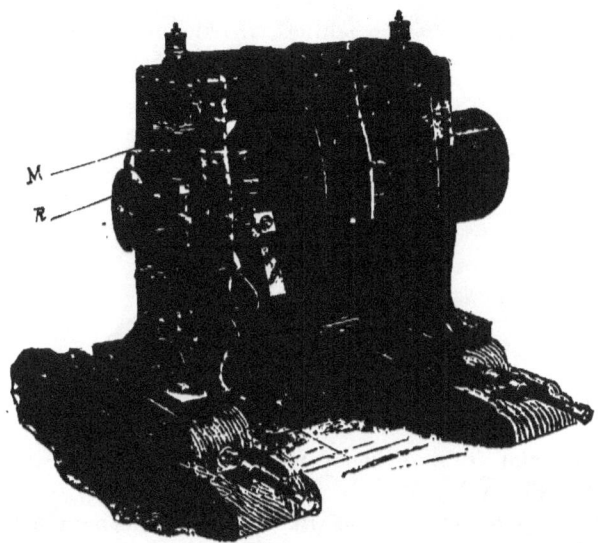

FIG. 2.—CONTINUOUS CONSTANT-CURRENT BIPOLAR GENERATOR.

brushes back or forward over the commutator, under the influence of an electromagnet M.

Constant-current machines are made for as many as 200 arc lights; *i. e.*, about 10,000 volts and 9 amperes, or an output up to 90 kilowatts capacity, but such large sizes are exceptional.

15. Constant-potential machines may be subdivided into sub-classes, according to the arrangement for supplying their magnetic flux—namely:

(a.) *Magneto-electric* machines, in which permanent magnets are employed for the fields.

The magneto-electric generator was the original type and progenitor of the dynamo, or dynamo-electric generator—but has almost entirely disappeared. It is, however, still used in telephony, the hand call being a small alternating-current magneto generator, driven by power applied at a handle. The magneto-electric generator is also used in firing mining fuses, and in some signaling and electro-therapeutic apparatus.

Fig. 3 represents a form of magneto-electric generator. M, is a triple group of permanent magnets, and A, is the armature.

(b.) *Separately-excited* machines, in which the field electromagnets are excited by electric current from a separate electric source.

FIG. 3.—ALTERNATING-CURRENT MAGNETO-ELECTRIC GENERATOR.

A particular form of separately excited generator is represented in Fig. 4.

Here a generator A, has its field magnets supplied by a small generator B, employed for this sole purpose. It is not necessary, however, that the exciting machine be used exclusively for excitation. Thus two generators, each employed in supplying a load, and each supplying the field magnets of the other, would be mutually separately excited.

In central stations large continuous-current machines are occasionally, and alternating-current machines are usually, separately excited.

(c.) *Self-excited* machines, or generators whose field magnets are supplied by currents from the armature.

Fig. 5 represents a form of self-excited generator. M, M, are the field magnets, P, the *pilot lamp;* i. e., a lamp connected across the terminals of the machine, to show that the generator is at work. S, the main circuit switch, R, the rocker-arm carrying the brushes B, B.

16. Self-excited machines may be divided into three classes; viz.,

(1.) Series wound.
(2.) Shunt wound.
(3.) Compound wound.

Series-wound machines have their field magnets connected in series with their armatures. The field winding consists of

FIG. 4.—ALTERNATING-CURRENT MULTIPOLAR SEPARATELY-EXCITED GENERATOR.

stout wire, in comparatively few turns. Arc-light machines are almost always series wound. Fig. 6 represents a particular form of series-wound machine for arc-light circuits. Here the current from the armature passes round the cylindrical magnets M, M, through the regulating magnet m, and thence to the external circuit. The machine in Fig. 2 is also series wound.

Shunt-wound machines have their field magnets connected to the main terminals, that is, placed in shunt with the external circuit. In order to employ only a small fraction of the total current from the armature for this purpose, the resistance of the field magnets is made many times higher than the resistance of the external circuit. This is accomplished by winding the magnets with many turns of fine wire, carefully insulated.

A particular form of shunt-wound machine is represented in Fig. 7.

Here the fine wire windings of the four magnets coils are supplied in one series through the connecting wires *W*, *W*, *W*,

FIG. 5.—SELF-EXCITED CONTINUOUS-CURRENT GENERATOR.

from the main terminals of the machine, one of which is shown at *M*. In order to regulate the strength of the exciting current through the magnet circuit, it is usual to insert a hand-regulating resistance box, called the *field regulating box*, in series with them.

(d.) *Compound-wound machines*. These are machines that are partly shunt wound and partly series wound.

It is found that when the load increases on a series-wound generator, it tends to increase the pressure at its terminals; *i.e.*, to raise its E. M. F. On the other hand, when the load increases on a shunt-wound generator, it tends to diminish the pressure at its terminals; *i. e.*, to lower its E. M. F. In order, therefore, to obtain good automatic regulation of pressure from a machine under all loads, these two tendencies are so directed as to cancel each other; this is accomplished by employing a winding that is partly shunt and partly series.

Fig. 8 represents a particular form of a compound-wound machine.

Here there are two spools placed side by side on each magnet-core, one of fine wire in the shunt circuit, carrying a current, and exciting the fields, even when no current is supplied externally by the machine, and the other of stout wire making

FIG. 6.—SELF-EXCITED SERIES-WOUND CONTINUOUS-CURRENT GENERATOR.

comparatively few turns. This is part of the series winding which carries the current to the external circuit. The excitation of the magnets from this winding, therefore, depends upon the current delivered by the machine; *i. e.*, upon its *load*.

Many generators for incandescent lamp circuits, as well as many generators for power circuits are compound wound.

17. Besides the preceding classes, dynamo-electric machines may be conveniently divided into other classes, according to a variety of circumstances; for example, they may be divided according to the number of magnetic poles in the field frame, as follows:

(a) *Bipolar machines*, or machines having only two magnetic field poles.

Bipolar machines may be subdivided, according to the number of separate magnetic circuits passing through the exciting

FIG. 7.—SELF-EXCITED SHUNT-WOUND CONTINUOUS-CURRENT GENERATOR.

coils, into *single-circuit bipolar, double-circuit bipolar machines*, and so on. Generally, however, modern bipolar machines are not constructed with more than two magnetic circuits. Figs. 1, 2, 3 represent bipolar machines. Of these, Fig. 1 possesses a single magnetic circuit, and Fig. 2 a double magnetic circuit.

(b) *Multipolar machines*, or machines having more than two magnetic poles.

Fig. 9 represents a multipolar, diphase alternator of many

poles. This machine was employed at the World's Columbian Exhibition.

18. Multipolar machines may be divided into the following sub-classes :

Quadripolar, or those having four poles.
Sextipolar, or those having six poles.
Octopolar, or those having eight poles.
Decipolar, or those having ten poles.

Beyond the number of ten poles, it is more convenient to omit the Latin prefix, and to characterise the machine by the

FIG. 8.—COMPOUND-WOUND CONTINUOUS-CURRENT GENERATOR.

number of poles, as, for example, a 14-pole, or 16-pole machine, etc.

Quadripolar machines are common. Fig. 10 shows a quadripolar machine. This machine has four brushes and is compound wound. It is designed to supply from 500 to 600 volts pressure at its brushes, and is surmounted by a group of six pilot lights in series.

Fig. 7 also represents a quadripolar generator.

Fig. 11 shows a form of continuous-current, self-exciting, compound-wound, sextipolar machine, arranged for direct connection to the main shaft of an engine. The machine is provided, as shown, with six collecting brushes.

Fig. 12 shows an alternating-current, self-exciting, octopolar generator for arc circuits. Although this machine is an *alternator ; i. e.*, supplies alternating currents, it, nevertheless, supplies its field-magnet coils in series with continuous currents from the commutator *C*, at one end of its shaft. The magnet *M*, forms an essential part of a short-circuiting device, whereby the machine is automatically short-circuited, on the external circuit becoming accidentally broken, in which case

FIG. 9.—ALTERNATING-CURRENT, 750-KILOWATT DIPHASE MULTIPOLAR GENERATOR.

the pressure generated by the machine might become so great as to endanger the insulation of the armature.

Fig. 13 shows a decipolar alternator, separately excited, and compensating. This machine is belt-driven, and it drives in turn a small dynamo *D*, employed for exciting the ten field magnets. The commutator, shown at *C*, is provided for the purpose of automatically increasing the pressure at the brushes of the machine with the load, so as to compensate for drop of pressure in the line or armature. In other words, the machine is compound-wound.

STRUCTURAL ELEMENTS. 19

As we have already observed, bipolar machines may be subdivided into classes according to the number of magnetic circuits passing through their exciting coils. In general, multipolar machines may be similarly classified. But, as usually constructed, there are as many independent magnetic circuits as there are poles. Thus, a quadripolar generator has

FIG. 10.—CONTINUOUS-CURRENT SELF-EXCITED COMPOUND-WOUND QUADRIPOLAR GENERATOR.

usually four magnetic circuits, a sextipolar six, and so on. In some cases, however, a double system of field magnets is provided, one on each side of the armature; in this case, the number of magnetic circuits may be double the number of poles.

19. In designing a continuous-current generator, the number of poles in the field is, to a certain degree, a matter of choice. In almost all cases, directly-coupled, continuous-current dynamos are multipolar, while belt-driven dynamos are frequently bipolar. Directly-coupled, continuous-current dynamos are usually multipolar machines, owing to the fact that, in order to conform with engine construction, they have to be made with a comparatively slow speed of rotation, and, since

FIG. 11.—CONTINUOUS-CURRENT SELF-EXCITED GENERATOR.

the E. M. F. generated depends upon the rate of cutting magnetic flux, if the speed of the conductor is decreased, the total amount of flux must be correspondingly increased. This necessitates a greater cross-section of iron in the field magnets in order to carry the flux, and this large amount of iron is most conveniently and effectively disposed in multiple magnetic circuits. To a certain extent the number of poles is arbitrary, but usually, in the United States, the greater the output of a direct-driven generator, the greater the number of poles.

In alternators, however, the case is different. Here, in order to conform with a given system of distribution, the frequency of alternation in the current is fixed, and, since the speed of revolution of the armature is determined within certain limits, by mechanical considerations, or by the speed of the driving engine, the number of poles is not open to choice, but is fixed by the two preceding considerations. In any alternator, the number of alternations of E. M. F. induced per revolution in the coils of its revolving armature, is equal to the number of

FIG. 12.—ALTERNATING-CURRENT SELF-EXCITED OCTOPOLAR GENERATOR.

poles. Consequently, an alternator producing a frequency of 133∼; that is a frequency of 133 complete periods or cycles per second, delivers 266 alternations from each coil, and its armature must, therefore, pass 266 poles per second.

20. Fig. 16 shows a 12-pole alternator. The wires *a, a*, are in circuit with the field magnets, and serve to carry the current which excites them, while the wires *b, b,* lead from the brushes.

21. Dynamo-electric machines may also be divided, according to their magnetic circuits, into the two following classes:

(a.) Those having *simple magnetic circuits* formed by a single core and winding.

FIG. 13.—ALTERNATING-CURRENT SEPARATELY-EXCITED DECIPOLAR COMPENSATING GENERATOR.

FIG. 14.—CONTINUOUS-CURRENT CONSEQUENT-POLE BIPOLAR SHUNT-WOUND GENERATOR.

(b.) Those having *consequent poles*, or poles formed by a double winding; that is, by the juxtaposition of two poles of the same name. Dynamo-electric machines belonging to the first class are shown in Figs. 1, 3 and 5. A type of machine belonging to the consequent-pole class is shown in Figs. 14 and 15. The poles are shown at N, N, and S, S, in each case, the field coils being so wound and excited as to produce consequent poles.

FIG. 15.—CONTINUOUS-CURRENT CONSEQUENT-POLE BIPOLAR GENERATOR.

22. Dynamo machines may also be classified according to the shape of the armature, as follows; namely,
 (a.) *Ring armatures.*
 (b.) *Cylinder* or *drum armatures.*
 (c.) *Disc armatures.*
 (d.) *Radial* or *pole armatures.*
 (e.) *Smooth-core armatures.*
 (f.) *Toothed-core armatures.*

Figs. 2 and 11 represent examples of ring armatures.

Since Gramme was the first to introduce the ring type of armature, this form is frequently called a *Gramme-ring armature.*

Figs. 1, 5 and 14, show examples of cylinder or drum armatures. Disc armatures are very seldom employed in the United States. An example of a disc armature is shown in Fig. 19. An example of a radial or pole armature is seen in Fig. 17.

A smooth-core armature is one on which the wire is wound over the cylindrical iron core, so as to cover the armature surface completely; or, if the wire does not cover the surface completely, the space between the wires may either be left vacant or filled with some non-magnetic metal. Such armatures are represented in Figs. 1, 2, 5, 15.

A toothed-core armature, on the other hand, is one on which

FIG. 16.—ALTERNATING-CURRENT SEPARATELY-EXCITED 12-POLE GENERATOR.

the wire is so wound in grooves or depressions, on the surface of the laminated iron core, that the finished armature presents an ironclad surface, but with slots containing insulated copper wire. Such an armature is shown in Fig. 18 and also in Figs. 7, 10 and 11. It is frequently called an *iron-clad armature*.

23. Dynamos may also be divided, according to the actual or relative movement of armature or field, into the following classes; namely,

(a.) Those in which the field is fixed and the armature

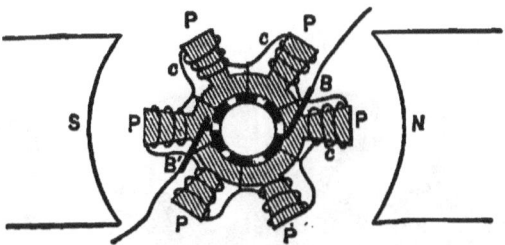

FIG. 17.—DIAGRAM OF POLE ARMATURE.

revolves. This class includes all the machines previously described, except that represented in Fig. 19.

(b.) Those in which the armature is fixed and the field revolves. An example of this type of machine is shown in

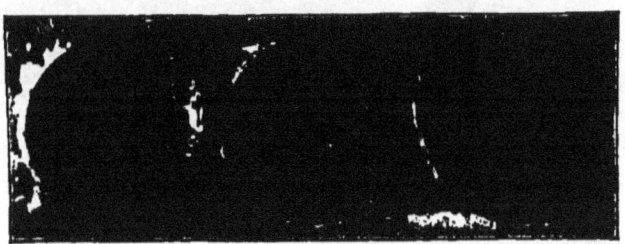

FIG. 18.—A TOOTHED-CORE ARMATURE SHOWING THE STAGES OF WINDING.

Fig. 19 A and B, where two sets of field magnets, mounted on a common shaft, revolve together around a fixed disc armature, shown in Fig. 19 B, which is rigidly supported vertically in the space between them.

(c.) Those in which the field and armature are both fixed, but the magnetic connection between the two is revolved. These dynamos are usually called *inductor dynamos*.

24. Dynamo machines may also be divided, according to the character of the work they are intended to perform, into the following classes; namely,

    (a.) *Arc-light generators.*
    (b.) *Incandescent-light generators.*
    (c.) *Plating generators.*
    (d.) *Generators for operating motors.*

FIG. 19A.—ALTERNATING-CURRENT DOUBLE 12-POLE GENERATOR WITH FIXED ARMATURE AND REVOLVING FIELD FRAMES.

    (e.) *Telegraphic generators.*
    (f.) *Therapeutic generators.*
    (g.) *Welding generators.*

25. Alternating-current generators may be divided, according to the number of separate alternating currents furnished by the machine, into the following classes; namely,

    (a.) *Uniphase alternators,* or those that deliver a single alternating current. To this class of machines belong all the ordinary alternators employed for electric lighting purposes.

    (b.) *Multiphase alternators,* or those that deliver two or more alternating currents which are not in step.

Some multiphase alternators can supply both single-phase and multiphase currents to different circuits.

Multiphase machines may be further subdivided into the following classes; namely,

(1.) *Diphase machines*, or those delivering two separate alternating currents. These two currents are, in almost all cases,

FIG. 19B.—DISC ARMATURE.

quarter-phase currents, that is to say, they are separated by a quarter of a complete cycle. Although it is possible to employ any other difference of phase between two currents, yet the quarter-phase is in present practice nearly always employed.

Fig. 9 represents a diphase generator, or *diphaser*.

(2.) *Triphase machines*, or *triphasers*, are generators delivering three separate alternating currents. These three currents are, in all cases, separated by one third of a complete cycle.

Uniphase machines are sometimes called *single-phase machines*, and diphase machines are sometimes called *two-phase machines* or *two-phasers*, while triphase machines are sometimes called *three-phase machines* or *three-phasers*. The terminology above employed, however, is to be preferred.

26. In addition to the above classification there are the following outstanding types:

(a.) *Single-field-coil multipolar machines*, or machines in which multipolar magnets are operated by a single exciting field coil.

(b.) *Commutatorless continuous-current machines*, or so-called *unipolar machines*, in which the E. M. Fs. generated in the armature, being obtained by the continuous cutting of flux in a uniform field, have always the same direction in the circuit, and do not, therefore, need commutation. The term unipolar is both inaccurate and misleading, as a single magnetic pole does not exist.

CHAPTER III.

MAGNETIC FLUX.

27. A magnet is invariably accompanied by an activity in the space or region surrounding it. Every magnet produces a magnetic field or flux, which not only passes through the substance of the magnet itself, but also pervades the space surrounding it. In other words, the property ordinarily called *magnetism* is in reality a peculiar activity in the surrounding ether, known technically as *magnetic flux.*

By a simple convention magnetic flux is regarded as passing out of the north-seeking pole of a magnet, traversing the space surrounding the magnet, and finally re-entering the magnet at its south-seeking pole. Magnetic flux, or magnetism, is circuital; that is, the flux is active along closed, re-entrant curves.

28. Although we are ignorant of the true nature of magnetic flux, yet, perhaps, the most satisfactory working conception we can form concerning it, is that of the ether in translatory motion; in other words, in a magnet, the ether is actually streaming out from the north-seeking pole and re-entering at the south-seeking pole.

Since the ether is assumed to possess the properties of a perfect fluid; *i. e.*, to be incompressible, readily movable, and almost infinitesimally divisible, it is evident that if a hollow tube, or bundle of hollow tubes, of the same aggregate dimensions as a magnet, be conceived to be provided internally with a force pump in each tube, and that such tube be placed in free ether, then, on the action of the force pumps, a streaming would occur, whereby the ether would escape from one end of each of the tubes, traverse the surrounding space, and re-enter at the other ends of the tubes. Moreover, if the stream lines, through which the ether particles would move under such ideal circumstances, were mapped out, they would be found to coincide with the observed paths which the magnetic stream lines take in the case of a magnet.

Similar stream lines could be actually observed in the case of a hollow tube provided internally with a pump, and filled with and surrounded by water; only, in this case, on account of the friction of the liquid particles, both in the tube and between themselves, work would require to be done and energy expended in maintaining the motion, and, unless such energy

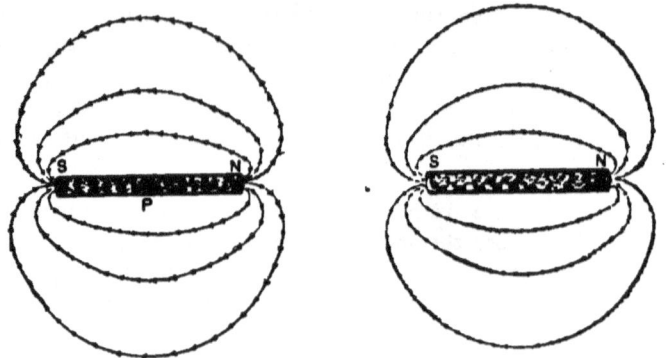

FIGS. 20 AND 20A.—DIAGRAMS REPRESENTING A TUBE, IMMERSED IN WATER, WITH A FORCE-PUMP AT ITS CENTRE AND HYDROSTATIC STREAM LINES — AND A CYLINDRICAL BAR OF IRON, MAGNETIZED, I. E., WITH A M. M. F. ACTIVE WITHIN IT AND MAGNETIC FLUX STREAM LINES.

were supplied, the motion would soon cease. In the case of the ether, however, there being, by hypothesis, no friction, although energy would probably be required to set up the motion, yet, when once set up, no energy would be required to sustain it, and the motion should continue indefinitely. This is similar to what we find in the case of an actual steel magnet. The above theory is merely tentative. The real nature of magnetism may be quite different; but, for practical purposes, assuming its correctness, since there is no knowledge as to the pole of the magnet from which the ether issues, it is assumed, as above stated, to issue from the north-seeking pole.

29. Fig. 20 represents, diagramatically, a tube provided at its centre with a rotary pump P, and immersed in water. If the pump were driven so as to force the water through the tube in the direction of the arrows; *i. e.*, causing the water to enter the tube at *S*, and leave it at *N*, then stream lines would be produced in the surrounding water, taking curved paths, some of which are roughly indicated by arrows.

Fig. 20A represents the application of this hypothesis to the case of a bar magnet of the same dimensions as the tube. Here the *magneto-motive force* of the magnet corresponds to the water-motive force of the pump in Fig. 20, and is hypothetically assumed to cause an ether stream to pass through the magnet in the direction indicated by the arrows ; namely, to enter the magnet at the south pole and issue at the north pole. These ether streams would constitute hypothetically the magnetic

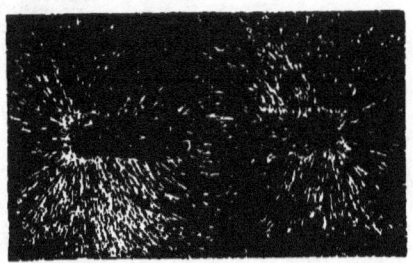

FIG. 21.—DISTRIBUTION OF FLUX ABOUT A STRAIGHT BAR MAGNET IN A HORIZONTAL PLANE, AS INDICATED BY IRON FILINGS.

flux, and would pass through the surrounding space in paths roughly indicated by the arrows. The actual flux paths that would exist in the case of a uniformly magnetized short bar magnet are more nearly shown in Fig. 21. Here it will be noticed that the flux by no means issues from one end only of the magnet, re-entering at the other end. On the contrary, the flux, as indicated by chains of iron filings, issues from the sides as well as from the ends of the bar. The reason for this is evidently to be found in the fact, that each of the particles or molecules of the iron, is, in all probability, a separate and independent magnet, and therefore must issue its own ether stream independently of all the rest. The effect is therefore not unlike that of a very great number of minute voltaic cells connected in series into a single battery, and the whole immersed in a conducting liquid where side leakage can exist.

30. The *magnetic field*, that is the space permeated by magnetic flux, may be mapped out in the case of any plane section by the use of iron filings. For example, Fig. 21, before alluded to, as representing the flux of a straight-bar magnet, had its flux paths mapped out as follows: A glass plate, covered with a thin layer of wax, was rested horizontally on a bar magnet, with its wax surface uppermost. It was then dusted over with iron filings and gently tapped, when the iron filings arranged themselves in chain-forms, which are approximately those of

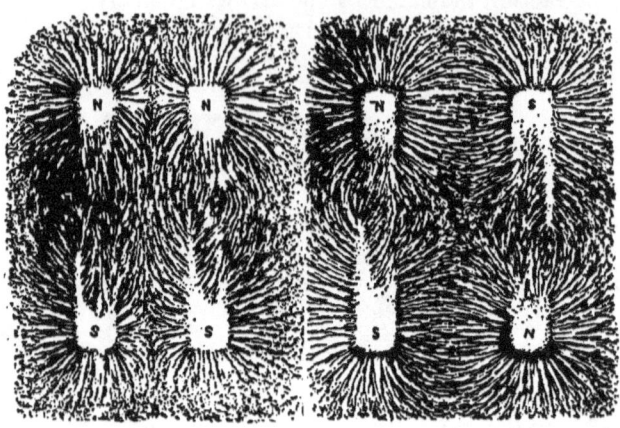

FIGS. 22, A AND B.—MAGNETIC FIELDS BETWEEN PARALLEL BAR MAGNETS.

the stream-lines of magnetic flux. A satisfactory distribution having been obtained in this manner, the glass plate was gently heated in order to fix the filings. On cooling, the filings were sufficiently adherent to the plate to permit it to be used as the positive from which a good negative picture can be readily obtained by photographic printing.

31. A modification of the above process was employed in the case of Figs. 22, A and B, shown above. Here a photographic positive was obtained by forming the field, in the manner previously explained, on a sensitized glass plate in a dark room, instead of on a waxed plate; and, after a satisfactory grouping of filings had been obtained under the influence of the field, exposing the plate momentarily to the action of light, as, for example, by the lighting of a match. The filings are then removed, the plate developed, and the negative so obtained employed for printing.

32. Magnetic flux may vary in three ways; namely,
(1.) In direction.
(2.) In intensity.
(3.) In distribution.

The direction of magnetic flux at any point can be readily determined by the direction assumed at that point, by the magnetic axis of a very small, delicately suspended compass needle. The compass needle always comes to rest as if threaded by the flux, which enters at its south pole, and leaves it at its north pole, thus causing the needle to point in the direction of the flux. Assuming that a compass needle may be represented

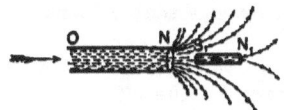

FIG. 23.—HYDRAULIC ANALOGUE SHOWING ATTRACTION OF OPPOSITE POLES.

by a little tube containing an ether force pump, the tube would evidently come to rest when the flux it produced passed through it in the same direction as the flux into which it was brought. That is to say, if the needle be brought into the neighborhood of a north pole, it will come to rest with its south pole pointing toward the north pole of the controlling magnet, since in this way only could a maximum free ether motion be obtained. If, however, the compass needle be held in the opposite direction; *i. e.*, with its north-seeking pole toward the north-seeking pole of the magnet, the two opposed stream lines will, by their reaction, produce a repellent force. These effects are generally expressed as follows:

Like magnetic poles repel, unlike magnetic poles attract. Strictly speaking, this statement is not correct, since, whatever theory of magnetism be adopted, it is the fluxes and not the poles which exercise attraction or repulsion.

33. Fig. 23 represents the action of the flux from a magnet upon a small compass needle, as illustrated by the hydraulic analogy. The water is represented as streaming through the tube $O\ N$, and issuing at the end N, in curved stream lines. Suppose the small magnet, or compass needle $S\ N$, also has a stream of water flowing through it, entering at S_1, and leaving at N_1. Then, if the compass needle be free to move about its centre of figure, it will come to rest when the stream from the large tube $O\ N$, flows through the smaller tube from S_1 to N_1 that is, in the direction of its own stream.

If, however, the small tube $S_1\ N_1$ is not free to move, but is fixed with its end N_1 toward the end N of the larger tube, as

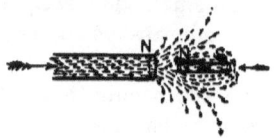

FIG. 24.—HYDRAULIC ANALOGUE SHOWING REPULSION OF SIMILAR POLES.

shown, in Fig. 24, then the opposite streams will conflict, and produce, by their reaction, the effect of repulsion between the tubes.

34. Magnetic flux possesses not only definite direction, but also magnitude at every point; that is to say, the flux is stronger nearer the magnet than remote from it. For example, considering a magnet as being represented by a tube with an ether force pump, the velocity of the ether flux will be a maximum inside the tube, and will diminish outside the tube as we recede from it. The intensity of magnetic flux is generally called its *magnetic intensity* or *flux density*.

Faraday, who first illustrated the properties of a magnetic field, proposed the term *lines of magnetic force*, and this term has been very generally employed. The term, however, is objectionable, especially when an attempt is made to conceive of magnetism as possessing flux density, or as varying in intensity at any point; for, in accordance with Faraday's conception, the idea of an increased flux would mean a greater number of lines of magnetic force traversing a given space. While this might be assumed as possible, still the conception that magnetism acts along lines, and not through spaces, is very misleading. An endeavor has been made to meet this objection by the introduction of the term *tubes of force*. A far simpler working conception is that of *velocity of ether*. that is, increased quantity passing per second, as suggested by the force-pump analogue. Here the increased flux density at any point would simply mean an increase of ether velocity at such point.

35. Intensity of magnetic flux is measured in the United States, in units called *gausses*, after a celebrated German magnetician named Gauss. A gauss is an intensity of one line of force, or unit of magnetic flux, per square centimetre of cross-sectional area, and is an intensity of the same order as that produced by the earth's magnetism on its surface. For example, the intensity of the earth's flux at Washington is about 0.6 gauss, with a dip or inclination of approximately 70°.

Magnetic flux may be *uniform* or *irregular*. Fig. 25 A, shows a uniform flux distribution, as represented diagrammatically, by straight lines at uniform distances apart. That is to say, uniform intensity at any point is characterized by rectangularity of direction in path at that point. Irregular intensity is characterized by bending, and the degree of departure from uniformity is measured by the amount of the bending. Irregular flux density may be either *converging*, as at B, or *diverging*, as at C. Convergent flux increases in intensity along its path, and divergent flux diminishes.

36. When the flux paths are parallel to one another, the intensity must remain uniform. Thus, in Fig. 25 at A, let the area, $ABCD$, be 1 square centimetre, then the amount of flux which passes through it in this position, or, in our hydraulic analogue, the quantity of water which would flow through it in a given time, will be the same if the area be shifted along the stream line parallel to itself into the position $EFGH$.

When the flux converges, as at B, in Fig. 25, then the amount of flux passing through the normal square centimetre $IJKL$, will, further on, pass through a smaller intercepting area, say one-fourth of a square centimetre $MNOP$, and consequently, the intensity at this area would be four times greater, and, in the hydraulic analogy, the same quantity of water passing per second, flowing through a cross sectional area four times more constricted, will flow there with four times the velocity.

When the flux diverges, as at C, the opposite effect is produced. Thus the flux shown in the figure as passing through the area $Q\ R\ S\ T$, say one-fourth of a square centimetre,

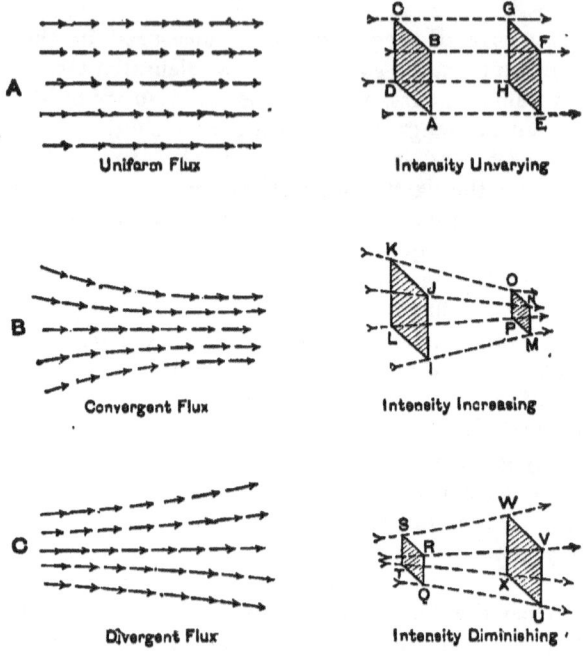

FIG. 25.—VARIETIES OF FLUX.

would, at $U\ V\ W\ X$, pass through one square centimetre, at four times less density, or, in the case of the hydraulic analogy, at one-fourth of the velocity.

37. The existence of a magnetic flux always necessitates the expenditure of energy to produce it. In the case of the ether pump, assuming that energy is required to establish the flow through the tube, this energy being imparted to the ether, becomes resident in its motion, so that ether, plus energy of motion, necessarily possesses different properties from ether at rest. In the same way in the case of a magnet, the energy required to set up the magnetic flux; *i. e.*, to magnetize it, is undoubtedly associated with such flux. Wherever the magnetic intensity is greatest, there the corresponding ether velocity, according to our working hypothesis, is greatest, and in that portion of space the energy of motion is greatest.

38. It is well known, dynamically, as a property of motion, that the energy of such motion in a given mass varies as the square of the velocity, so that, by analogy, if magnetic flux density corresponds to ether velocity, we should expect that the energy associated with magnetic flux should increase with

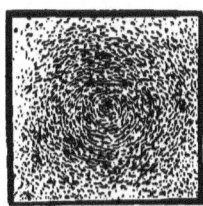

FIG. 26.—DISTRIBUTION OF FLUX ROUND A VERTICAL WIRE CARRYING A CURRENT, AS INDICATED BY IRON FILINGS.

the square of its intensity. This is experimentally found to be the case. Thus if $\mathcal{B}$, represents the intensity of magnetic flux, expressed in gausses, then the energy in every cubic centimetre of space, except in iron or other magnetic material; *i. e.*, in the ether permeated by such intensity, is $\frac{\mathcal{B}^2}{8\pi}$ ergs. Thus, if a cubic inch of air (a volume of 16.387 cubic centimetres), be magnetized to the intensity of 3,000 gausses, the energy it contains, owing to its magnetism, will be

$$\frac{16.387 \times 3,000 \times 3,000}{8 \times 3.1416} = 0.5868 \times 10^7 \text{ ergs.} = 0.5868 \text{ joule.}$$

39. Just as in the electric circuit, the presence of an electric current necessitates the existence of an E. M. F. producing it, so in a magnetic circuit, the presence of a magnetic flux necessitates the existence of a magneto-motive force (M. M. F.) producing it.

We know of but two methods by which a M. M. F. can be produced, viz.:

(1.) By the passage of an electric current, the neighborhood of which is invested with magnetic properties; *i. e.*, surrounded by magnetic flux;

(2.) As a property inherent in the ultimate particles of cer-

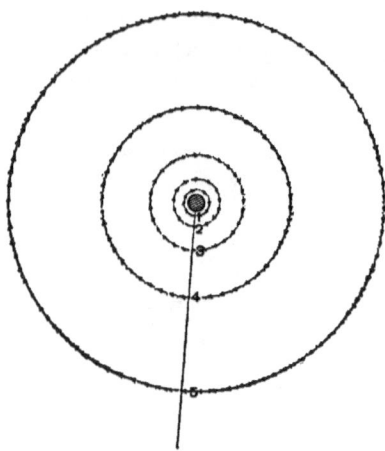

FIG. 27.—GEOMETRICAL DISTRIBUTION OF FLUX PATHS ROUND A WIRE CARRYING A CURRENT.

tain kinds of matter, possibly the molecules, of the so-called magnetic metals.

The passage of an electric current through a long, rectilinear conductor, is attended by the production of a magnetic field in the space surrounding the conductor. The distribution of flux in this field, is a system of cylinders concentric to the conductor, and is directed clock-wise around the conductor, if the current be supposed to flow through the clock from its face to its back. This distribution is shown in Figs. 26, 27 and 28. Fig. 26 represents the distribution as obtained by iron filings. The density of the flux is roughly indicated by the density of the corresponding circles.

40. Fig. 27 shows the geometrical distribution of the flux paths around a wire carrying a current, which is supposed to flow from the observer through the paper. Here a few of the flux paths are indicated by the circles, 1, 2, 3, 4 and 5, while the direction is shown by the arrows. The distribution of the flux is such that it varies in intensity, outside the wire, inversely as the distance from the axis of the wire, and the total flux between any adjacent pair of circles in the figure is the same,

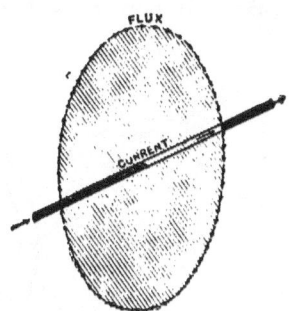

FIG. 28.—DIAGRAM OF RELATIVE DIRECTIONS OF MAGNETIC FLUX AND ELECTRIC CURRENT.

for example, between 1 and 2, or between 4 and 5. Or, in the hydraulic analogue, the total flow of water per second, between any pair of adjacent circles is the same, as between the circles 2, 3, or 4, 5, the velocity diminishing as the distance from the axis of the wire.

Fig. 28 represents the direction of the flux round the active conductor, the current flowing from the observer through the shaded disc.

41. The physical mechanism of the magnetic flux produced by a current is unknown, but if an electric current be assumed to be due to a vortex motion of ether in the active wire, the direction of which is dependent on the direction of the current through the wire, then such vortex motion will be accompanied by such a distribution of circular stream-lines in the ether, as is actually manifested, and, when the direction of the current through the conductor is changed, the direction of the stream-lines outside the conductor will also necessarily be changed.

As the strength of the current through the wire increases, the velocity of the ether surrounding the wire increases; *i. e.*, the intensity of the magnetic field everywhere increases.

42. If a conductor conveying a current be bent in the form of a circle as shown in Fig. 29, and a current, of say one ampere, be sent through the conductor, there passes through the loop so formed a certain number of stream-lines as represented diagrammatically. If now, the current in the wire be

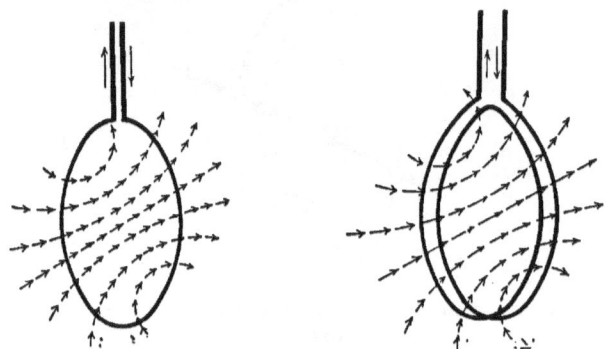

FIGS. 29 AND 30.—SINGLE LOOP OF ACTIVE CONDUCTOR, THREADED WITH FLUX, AND DOUBLE LOOP WITH M. M. F. DOUBLED.

doubled, that is increased to two amperes, the flux intensity everywhere will be doubled. The same effect, however, can be practically obtained by sending one ampere through the double loop, shown in Fig. 30, provided the two turns lie very close together. Magnetic flux through a loop, will depend, therefore, upon the number of ampere-turns, so that, by winding the loop in a coil of many turns, the flux produced by a single ampere through the coil may be very great. The M. M. F. produced by a current, therefore, depends upon the number of ampere-turns.

43. The *unit of M. M. F.* may be taken as the *ampere-turn*, and it frequently is so taken for purposes of convenience. The fundamental unit, however, of M. M. F., in the United States, is the *gilbert*, named after one of the earliest magneticians, Dr. Gilbert, of Colchester. The gilbert is produced by $\frac{1}{4\pi}$ of a

C. G. S. unit current-turn, and, since the C. G. S. unit of current is ten amperes, the gilbert is produced by $\frac{10}{4\pi}$ ampere-turn (0.8 approximately, more nearly 0.7958). It is only necessary, therefore, to divide the number of ampere-turns in any coil of

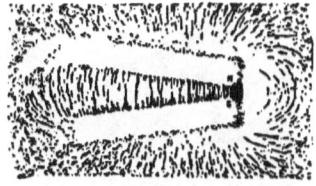

FIG 31.—DISTRIBUTION OF FLUX IN PLANE OVER A HORSE-SHOE MAGNET.

wire by 0.8, that is to multiply the number of ampere-turns by 1.25, more nearly 1.257) to obtain the M. M. F. of that coil expressed in gilberts.

44. Figs. 31 to 42 are taken from actual flux distributions as obtained by iron filings, and represent a series of negatives or positives secured by the means already described. A study of

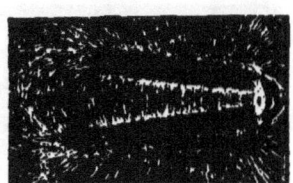

FIG. 32.—DISTRIBUTION OF FLUX IN PLANE OVER A HORSE-SHOE MAGNET.

such flux-paths assists the student to mentally picture the flux distributions which occur in practice.

Figs. 31 and 32 are the respective positive and negative photographic prints taken in the case of a horse-shoe magnet. Here the filings are absent in a region outside the magnet in the neighborhood of the poles $N S$. The cause of this is as follows: the fields were obtained by sprinkling iron filings over a smooth glass surface; the tapping of the surface necessary to insure the arrangement of the filings under the influence of the magnetic flux, has caused an accumulation of filings around these poles at the expense of the gap immediately in front of the poles which would otherwise be more fairly filled.

FIG. 33.—DISTRIBUTION OF FLUX BY IRON FILINGS IN PLANE OVER POLES OF ELECTRO-MAGNET.

45. The student should carefully avoid being misled by the supposition that the relative attractive tendencies of the iron filings in such diagrams represent the corresponding densities of the magnetic flux, for the reason that in a uniform magnetic flux such as shown at A, in Fig. 25, there is no attraction of iron filings, whatever its intensity, although, of course,

FIG. 34.—DISTRIBUTION OF FLUX BY CUT IRON WIRE IN PLANE OVER POLES OF ELECTRO-MAGNET.

a directive tendency still exists. In order that there should be any attractive tendency, in contradistinction to a mere directive tendency, it is necessary that the intensity of the magnetic flux shall vary from point to point; or, in other words, that the flux shall be convergent. The greater the degree of convergence the greater the attractive force. Consequently, variations of flux intensity as indicated by iron filings always exaggerate the appearance of flux density. Generally speaking, it is only the directions assumed by the filings in such diagrams, as indicative of the directions of the flux, which can be regarded as trustworthy. The neglect of this consideration has given rise to a popular belief that magnetic streamings occur with greater density at points, than at plane or blunt surfaces, which is not the case. There must necessarily be a rapid convergence or divergence of mag-

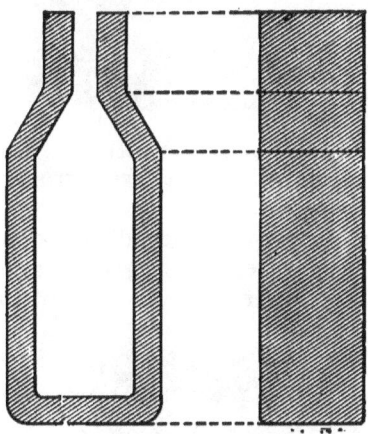

FIG., 35.—PLAN AND SIDE ELEVATION OF MAGNET EMPLOYED IN CONNECTION WITH FIGS. 36 AND 37.

netic flux at points, although the maximum density may not be very great. Owing to this convergence, iron filings, particles, nails, etc., are attracted more powerfully at such points, even though the uniform intensity of flux at plane surfaces in the vicinity may be greater.

46. Fig. 33 shows the distribution of magnetic flux as obtained by iron filings in a horizontal plane over the vertical poles of an electro-magnet. Here the flux-paths pass in straight lines between the nearest points of the adjacent poles, and in curved lines over all other parts of the plane. If we imagine, following the hydraulic analogue, that water streams proceed from minute apertures in one of the poles, and that

the magnet is immersed in water, then the stream-lines so produced in the water as it emerges from pole N, and enters through pole S, will be the same as is indicated by the iron

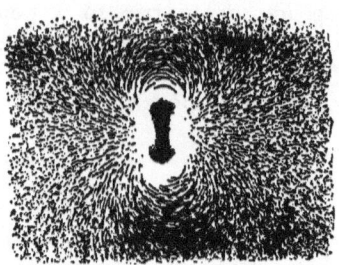

FIG. 36.—DISTRIBUTION OF FLUX BY IRON FILINGS IN PLANE OVER MAGNET SHOWN IN FIG. 35, WITH MAGNET PRESENTED VERTICALLY.

filings. Fig. 34 shows a similar distribution of flux over the poles of the same electro-magnet, where short pieces of fine soft iron wire were used in place of the iron filings.

FIG. 37.—DISTRIBUTION OF FLUX BY IRON FILINGS IN PLANE OVER MAGNET SHOWN IN FIG. 35, WITH MAGNET PRESENTED HORIZONTALLY.

Here the flux-paths have practically the same distribution as in the preceding case.

Figs. 36 and 37 show the distribution of flux by iron filings in a horizontal plane over the poles of the magnet represented in Fig. 35, the magnet being presented vertically in Fig. 36, and horizontally in Fig. 37, to the plane. Here the general distribution of flux between the polar surfaces is rectilinear.

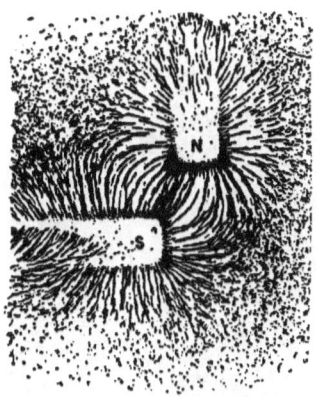

FIG. 38.—FLUX-PATHS BETWEEN DISSIMILAR POLES.

Fig. 38 illustrates the flux distribution attending the approach of what are called unlike poles. Here the ether

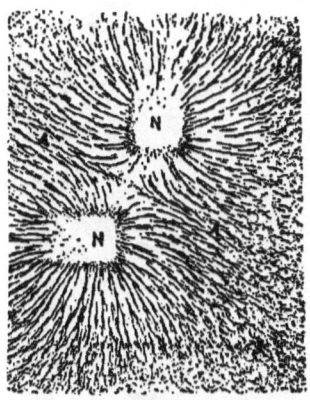

FIG. 39.—FLUX-PATHS BETWEEN SIMILAR POLES.

streams we assume to issue from N, in entering the magnet S, take the paths indicated.

Fig. 39 illustrates the flux distribution attending the

approach of what are called like poles. Here the hypothetical ether streams issuing from N, N, impinge, as shown, and pro-

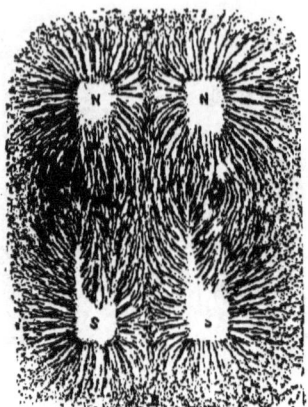

FIG. 40.—FLUX-PATHS BETWEEN TWO PARALLEL BAR MAGNETS, SIMILAR POLES ADJACENT.

duce a neutral line, $A A$, corresponding to *slack water* in the hydraulic analogue.

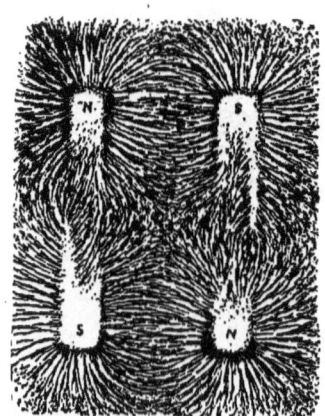

FIG. 41.—FLUX-PATHS BETWEEN TWO PARALLEL BAR MAGNETS OPPOSITE POLES ADJACENT.

Fig. 40 shows the distribution of flux in the case of two straight bar magnets laid side by side with like poles opposed.

The imaginary ether streams again oppose and the neutral line *B B*, is produced as shown.

Fig. 41 shows the distribution of magnetic flux in the case of two straight bar magnets, laid side by side, with unlike poles opposed. Here, according to hypothesis, some of the ether streams issuing from each magnet, pass back through the other magnet, the remainder closing their circuit through

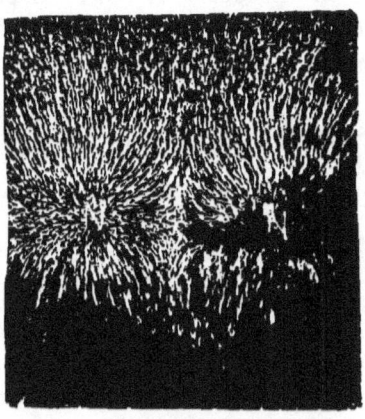

FIG. 42.—FLUX-PATHS SURROUNDING ANOMALOUS MAGNET.

the air outside. A curious central region between the magnets, bounded by curves resembling hyperbolas is shown at *C*, where, by symmetry, no ether motion penetrates, and thus corresponding, in the hydraulic analogue, to *calm water*.

Fig. 42 shows the distribution of flux over the surface of what is commonly called an *anomalous magnet*, that is a magnet having two similar poles united at its centre; or, in other words, having two separate magnetic circuits. Here the distribution of flux is similar to that in Fig. 40, where like poles are approached.

CHAPTER IV.

NON-FERRIC MAGNETIC CIRCUITS.

47. As we have already seen, magnetic flux always flows in closed paths, or forms what is called a *magnetic circuit*. The quantity of magnetic flux in a magnetic circuit depends not only upon the magneto-motive force, but also on the disposition and nature of the circuit. For example, it is not to be supposed that the flux produced by the 12 ampere-turns (15.084 gilberts) in the right-handed coil or helix of Fig. 43, by one ampere flowing through the twelve turns shown, would be

FIG. 43.—RIGHT-HANDED HELIX OF 12 TURNS CARRYING ONE AMPERE.

exactly the same, either in magnitude or distribution, as the flux from a single turn carrying 12 amperes, although the M. M. F. would be the same in each case. Just as in the case of an electric circuit, the current produced by a given E. M. F. depends on the resistance of the circuit, so in the case of a magnetic circuit, the magnetic flux produced by a given M. M. F. depends on a property of the circuit called its *magnetic reluctance*, or simply its *reluctance*.

Magnetic reluctance, therefore, is a property corresponding to electric resistance, and is sometimes defined as the resistance of a circuit to magnetic flux.

The resistance, in ohms, of any uniform wire forming portion of an electric circuit is equal to the *resistivity*, or *specific resistance*, of the wire, multiplied by the length of the wire, and divided by its cross-sectional area. In the same way, the reluctance, in oersteds, of any uniform portion of a magnetic circuit, is equal to the *reluctivity*, or *specific magnetic resistance* of the portion, multiplied by its length in centimetres, and divided by its cross sectional area in square centimetres. The reluctivity of air, wood, copper, glass, and practically all substances except iron, steel, nickel and cobalt, is unity. Strictly speaking, the reluctivity of the ether in vacuous space is unity, but the difference between the reluctivity of vacuum and of all non-magnetic materials is, for all practical purposes, negligibly small. Thus, the reluctance of a cylinder of air space of 10 cms. long and 2 sq. cms. in cross-sectional area, is 5 oersteds.

48. The reluctance of a circuit is measured in *units of reluctance* called *oersteds*. An oersted is equal to the reluctance of a cubic centimetre of air (or, strictly speaking, of air-pump vacuum) measured between opposed faces.

Having given the reluctance of a magnetic circuit, and its total M. M. F., the flux in the circuit is determined in accordance with Ohm's law; that is $\Phi = \dfrac{\mathcal{F}}{\mathcal{R}}$ where Φ, is the flux in webers, $\mathcal{F}$, is the magneto-motive force in gilberts, and $\mathcal{R}$, the reluctance in oersteds. It may afford assistance to contrast the well-known expression: amperes $= \dfrac{\text{volts}}{\text{ohms}}$, with the corresponding magnetic expression, webers $= \dfrac{\text{gilberts}}{\text{oersteds}}$.

49. The *unit of magnetic flux*, in the United States, is called the *weber*, and is equal to the flux which is produced by a M. M. F. of one gilbert acting through a reluctance of one oersted, corresponding in the above expression to the *ampere*, the *unit of electric flux*, which is the electric flux or current produced by an E. M. F. of one *volt* through a resistance of one *ohm*. For example, if an anchor ring of wood, such as is represented in Fig. 44, have a cross section of 10 sq. cms. and be uniformly wrapped with insulated wire, then when the current passes through the winding, the magnetic circuit will be entirely confined to the interior of the coil or solenoid, and no magnetic flux will be perceptible in the region outside it. This is the only known form of magnetic circuit in which the flux-paths can be confined to a given channel. These flux-paths are all circular, and possess the same intensity around each circle. If the mean circumference of the ring be 60 cms., the reluctance of the magnetic circuit will be approximately $\dfrac{60}{10} =$

6 oersteds, as in the similar case of electric resistance. If the number of turns in the winding be 200, and the exciting current steadily maintained at four amperes, the M. M. F. in the magnetic circuit will be 800 ampere-turns, or 1,005.6 gilberts. From this the total flux through the ring will be $\frac{1,005.6}{6} = 167.6$ webers.

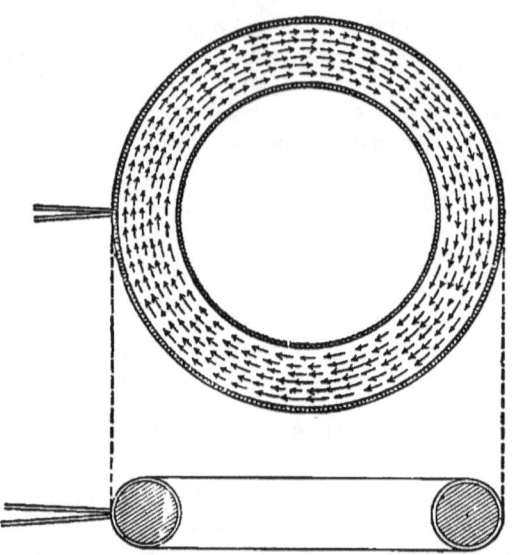

FIG. 44.—SECTIONS OF WOODEN RING UNIFORMLY WRAPPED WITH INSULATED WIRE CARRYING A CURRENT.

50. Besides the case of the anchor ring, represented in Fig. 44, the magnetic circuit of which, being entirely confined to the interior of the coil, permits its reluctance to be readily calculated, and the flux to be thus arrived at, another case, almost as simple, is afforded by a long straight helix of length l cms., uniformly wrapped with n, turns per cm. or $N = l\,n$, turns in all. Such a helix, when excited by a current of I amperes, develops a M. M. F. of $n\,I$ ampere-turns, or $1.257\,n\,I$ gilberts in each centimetre, or $1.257\,N\,I$ gilberts, for the total M. M. F.

The magnetic circuit of such a solenoid is roughly represented in Fig. 20 A. An inspection of this figure will show that flux passes through the interior of the helix in parallel streams, until it reaches a comparatively short distance from the ends, when it begins to sensibly diverge, and, emerging into the surrounding space, is diffused through widely divergent paths. That is to say, the magnetic circuit is characterized by two distinct regions; namely, that within the coil, where the flux is uniform, and, except near the ends, of a maximum intensity, and that outside and beyond the ends of the coil, where the flux is divergent and greatly weakened in intensity.

51. In the case of a long, straight, uniformly-wrapped helix, the reluctance of the circuit may be considered as consisting of two distinct portions; namely, a straight portion occupying the interior of the coil and lying practically between the ends, and a curved or diffused portion exterior to the coil. The reluctance of the first, or interior portion, will be practically $\dfrac{l}{a}$ oersteds, where a, is the cross sectional area of the interior of the coil in square cms. and l, the length of the coil in cms., or, more nearly, the reduced length of the non-divergent flux. It will be seen, therefore, that the interior of the coil behaves like a straight wire carrying electric flux, since it practically confines the flux to its interior, and, this particular portion of the magnetic circuit is similar to the case of the anchor ring above referred to, where the magnetic flux is confined to the interior of the ring.

Since the external circuit is diffused, its reluctance cannot be so simply expressed. Its value, however, may obviously be dealt with as follows: although the mean length of the flux-paths outside the coil is greater than in the interior portion, yet the area of cross section of the circuit is enormously extended. It would appear, therefore, that in the case of an indefinitely long straight coil, the external reluctance becomes negligibly small compared with the internal reluctance, and may be left out of consideration. In such a case, therefore, the flux established becomes

$$\Phi = \frac{1.257\, l\, n\, I}{\dfrac{l}{a}} = 1.257\, n\, I\, a \qquad \text{webers};$$

and, since, within the coil, this flux passes through a cross sectional area of a square centimeters, the interior intensity will be

$$\mathfrak{B} = \frac{1.257\, n\, I\, a}{a} = 1.257\, n\, I \qquad \text{gausses.}$$

Strictly speaking, therefore, this is the intensity of flux within an indefinitely long straight helix, and is approximately the intensity within helices which have lengths more than 20 times their diameter.

52. We have now discussed two cases of non-ferric circuits, whose reluctance is readily calculated; namely, a closed circular coil and a long straight helix.

In all other cases, the reluctance of a magnetic circuit is much more difficult to compute, although the fundamental relations remain unchanged.

When the magnetic circuit is non-ferric, although the reluctivity of the circuit always equals unity, yet, owing to the difficulty of determining the exact paths followed by the divergent flux, the reluctance is difficult to determine.

Most practical magnetic circuits, however, are composed either entirely, or mainly, of iron. At first sight, the introduction of iron into the circuit would appear to make the reluctance more difficult to determine, because the reluctivity of iron not only varies greatly with different specimens, but also with its hardness, softness, annealing, and chemical composition. Moreover, the apparent reluctivity of iron varies markedly with the density of the flux passing through it. Iron, when magnetically saturated, possesses a reluctivity equal to that of air; while, as we have seen, at low intensities, the reluctivity is much smaller, and may be several thousand times smaller.

Since, however, ferric circuits, as ordinarily employed, practically confine their flux-paths to the substance of the iron, and, since the reluctance of the iron is so much less than the reluctance of the *alternative air path* outside, the air flux may usually be neglected. Even where, owing to the reluctance of the air gaps in the circuit, such as in the case of dynamos and motors, a considerable amount of *magnetic leakage* or *diffusion* may take place through the surrounding air, yet it is preferable to regard this leakage as a deviation from the iron circuit, which may be separately treated and taken into account, and that the flux passes principally through the iron. For these reasons, ferric or aero-ferric circuits, at least in their practical treatment, are simpler to determine and compute than non-ferric circuits, since, although their reluctivity is variable at different points, yet the geometrical outlines of the flux-paths can be regarded as limited, and the reluctance of these paths can be readily determined approximately.

53. *Magnetizing force* may be defined as the space rate at which the magnetic potential descends in a magnetic circuit. Since the *total fall of magnetic potential* is equal to the M. M. F. in the circuit, just as the total "drop" in a voltaic circuit is equal to its E. M. F. Consequently, the line integral or sum of magnetizing force in a magnetic circuit must be equal to the M. M. F. in that circuit. In other words, if we multiply the rate of descent in potential by the distance through which that rate extends, and sum all such stages, we arrive at the total descent of magnetic potential. For instance, in Fig. 44 the total difference of magnetic potential is 1,005.6 gilberts, which, by symmetry, is uniformly distributed round the entire circuit. Since the mean length of this circuit is 60 cms. the rate of fall of potential is $\frac{1,005.6}{60} = 16.76$ gilberts-per-centimetre all round the ring, and this is, therefore, the magnetizing force, or, as it is sometimes called, the *magnetic force*. This magnetizing force is usually represented by the symbol $\mathcal{H}$, and, when no iron or magnetic metal is included in the circuit, is numerically identical with the flux density $\mathcal{B}$, so that $\mathcal{H}$, is expressed in gilberts-per-centimetre. The term magnetizing force was adopted from the old conception of magnetic poles; for, if a pole of unit strength could be introduced into a flux of intensity $\mathcal{H}$ gausses, the *mechanical force* exerted upon the pole would be $\mathcal{H}$ dynes, directed along the flux-paths. In any magnetic circuit, if we divide the M. M. F. in gilberts, by the length of a flux-path, we obtain the average value of the magnetizing force (or flux density in the absence of iron). Thus, in Fig. 21, if the long

helix there represented, has a M.M.F. of 5,000 gilberts, and a particular flux-path has a length of 500 cms., the mean magnetizing force, will be $\frac{5,000}{500} =$ 10 gilberts-per-centimetre, and the mean flux density will be 10 gausses, if there is no iron in the circuit. If there is iron, the mean *prime flux density* or *magnetizing force*, will still be 10 gilberts-per-centimetre, but the flux density established in the circuit will be greatly in excess of 10 gausses.

CHAPTER V.

FERRIC MAGNETIC CIRCUITS.

54. We will now proceed to study the phenomena which occur when iron is introduced into a magnetic circuit, as for example, into the circuit of the closed circular coil shown in Fig. 44, the mean interior circumference of which is 60 cms., and the mean cross sectional area 10 sq. cms. We have seen that if this ring be excited with 800 ampere-turns, or 1005.6 gilberts, the flux through the ring will be 167.6 webers; or, since the cross section of the ring is ten square centimetres, the intensity will be $\frac{167.6}{10} = 16.76$ gausses, and this intensity would remain practically unchanged if the substance of the ring were copper, brass, lead, zinc, wood, glass, etc. When, however, the ring is made of iron or steel, a very marked change takes place; the flux instead of being 167.6 webers, becomes, say, 170,000 webers, with a corresponding increase in intensity. This increase of flux in the circuit must either be due to an increase in the M. M. F., or to a diminution in the reluctance. It is usual to consider that iron conducts magnetic flux better than air; or, in other words, has a greater *magnetic permeability* than air. This idea corresponds to a reduction of reluctance similar to the reduction of resistance in an electric circuit. Although generally accepted, this conception is manifestly incorrect; for if the increased flux, due to the presence of iron in the ring, disappeared immediately on the removal of the M. M. F., there would be no preponderance of evidence in favor of either hypothesis. But the magnetic flux does not entirely disappear on the cessation of the prime M. M. F. On the contrary, in the case of a closed iron ring, the greater portion of the flux remains in the condition called *residual magnetism*.

55. It is evident, therefore, since M. M. F. is necessary to maintain the residual magnetic flux in the iron, that this

M. M. F. is the cause of the increase in magnetic flux when the prime M. M. F. is applied, and that, therefore, the increased flux cannot be due, except, perhaps, in a very small degree, to any change in the reluctivity of the medium, but to the establishment of a M. M. F. in the iron itself under the influence of the magnetizing flux. It is now almost certain that the ultimate particles of the iron, the molecules, or the atoms, are all initially magnets ; *i. e.*, inherently possess M. M. Fs. and magnetic circuits. The origin of this *molecular magnetism* in iron is, however, not yet known. In the natural condition, all the separate magnets of which iron is composed, are distributed indifferently in all directions, so that their circuits neutralize one another and produce no appreciable external effects. Under the influence of a magnetizing flux, these *molecular magnets* tend to become aligned, and to break up their original groupings. As they become aligned, and their M. M. Fs. become similarly directed, they are placed in series, and their effects are rendered cumulative, so that they exercise an increasing external influence, and an extending external flux. Or, taking the hydraulic analogue already referred to, and regarding each separate molecular magnet as a minute ether pump, as all the ether pumps are brought into line, the streams they are able to direct are increased in velocity, and are, therefore, carried further into the surrounding space. Consequently, the flux produced in the magnetic ring shown in Fig. 44, when furnished with an iron core, may be regarded as arising from two distinct sources of M. M. F. ; namely,

(1.) The *prime M. M. F.*, or that due to the magnetizing current which produces the flux through the circuit and substance of the iron, the value of which is practically the same as though the core were of wood or other non-magnetic material. This flux may be called the *prime flux* and possesses a corresponding prime intensity. In the case considered, the prime intensity or magnetizing flux density is 16.76 gausses. This magnetic intensity, acting upon the molecules of the iron, produces:

(2.) The *induced M. M. F.*, which may be called the *aligned* or *structural M. M. F.*, and depends for its magnitude not only upon the quality of the iron, but also upon the intensity of the prime flux. The harder the iron, and the greater its mechanical tendency to resist molecular distortion, the greater must be the prime intensity or the magnetic distorting power, in order to bring about the full structural M. M. F. When the prime intensity has reached such a magnitude that all the separate molecular magnets in the iron are similarly aligned, the iron is said to be *saturated*, and the M. M. F. it produces is a maximum, and, on the removal of the prime M. M. F. the structural M. M. F. will, in the case of a closed ring, largely remain, especially if the ring be of hard iron or steel. If, on

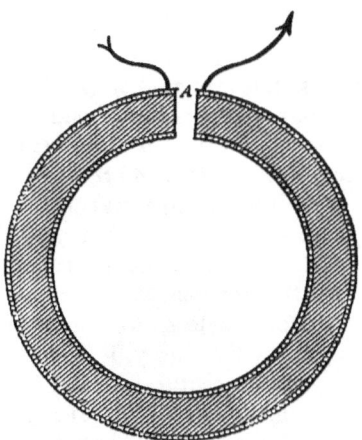

FIG. 45.—IRON RING PROVIDED WITH AIR-GAP, AND WOUND WITH WIRE.

the contrary, the ring be of soft iron, and have an air-gap cut in it, the structural M. M. F. may largely disappear. The relation between the structural M. M. F. and its flux, and the prime M. M. F. and the intensity which produces it, is complex, and can only be ascertained by experimental observation.

56. Fig. 45 represents the same iron ring with a saw-cut or air-gap at A, having a width of 0.5 cm. The reluctance of this air-gap, which, neglecting diffusion, has a length of 0.5 cms. and a cross-section of 10 sq. cms. is $\frac{0.5}{10} = 0.05$ oersted. If the total structural M. M. F., established in the ring under excitation, be 180,000 gilberts, then, immediately on the withdrawal of the prime M. M. F., the residual flux through the circuit will be $\frac{180,000}{6} = 30,000$ webers. Where this flux passes through the reluctance of the air-gap there will be established a C. M. M. F., just as in the electric circuit where a current of I amperes passes through a resistance of R, ohms, there is established a C. E. M. F. of IR volts. So that the C. M. M. F. has in this case the value, $F = \Phi R = 30,000 \times 0.05 = 1,500$ gilberts. This C. M. M. F. represents a mean demagnetizing force of $\frac{1,500}{60} = 25$ gilberts-per-centimetre, through the iron circuit. If this intensity of demagnetizing force is sufficient to disrupt the structural alignment of the molecular magnets, the residual magnetism will disappear. If, however, the intensity be less than that which the hardness of the iron requires to break up its structure, the residual magnetism will be semi-permanent.

Even though it be admitted that the preceding represents the true condition of affairs, and though it is the only existing hypothesis by which the phenomena of residual magnetism can be accounted for, nevertheless, for practical computations connected with dynamo machinery, it is more convenient to assume that there is no structural M. M. F. in iron, and that the difference in the amount of flux produced in ferric circuits is a consequence of decreased reluctance in the iron; or, in other words, that iron is a better conductor of magnetism. We will, therefore, in future, adopt the untrue but more convenient hypothesis.

57. The reluctivity of iron may be as low as 0.0005, but varies with the flux density; that is to say, the reluctance of a cubic centimetre of iron, measured between parallel faces, may be as low as 0.0005 oersted.

58. The fact has been established by observation, that in the magnetic metals, within the limits of observational error, a linear relation exists between reluctivity and magnetizing force. That is to say, within certain limits, as the magnetizing force brought to bear upon a magnetic metal increases, the apparent reluctivity of the metal increases in direct proportion. Thus, taking the case of soft Norway iron, its reluctivity, at a magnetizing force of 4 gilberts-per-centimetre, or prime magnetic intensity of 4 gausses, may be stated as 0.0005. Increasing the magnetizing force, the reluctivity increases by 0.000,057

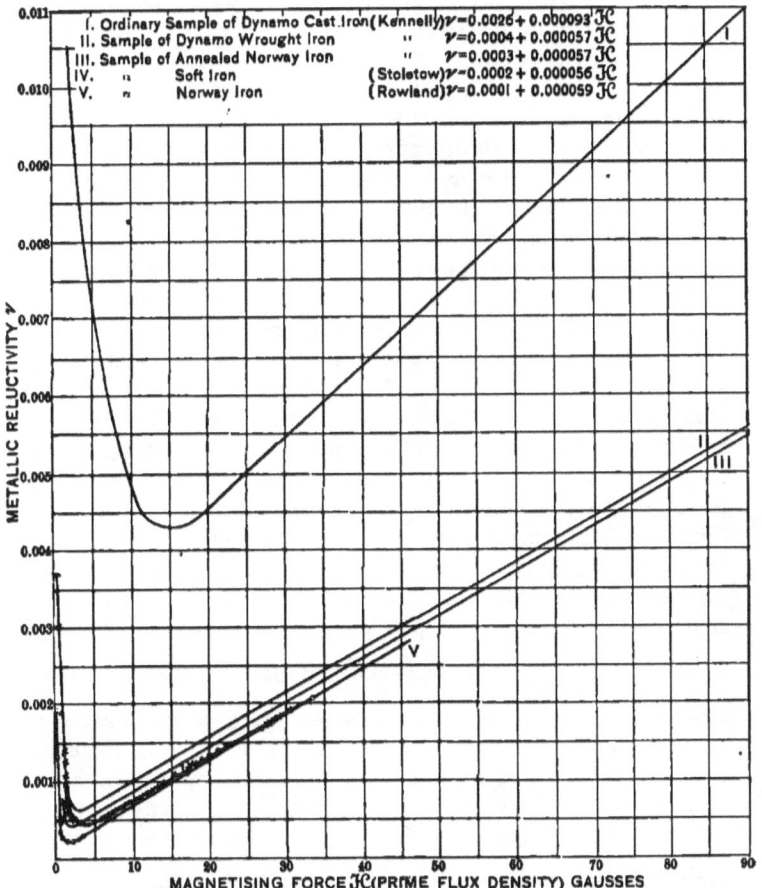

FIG. 46.—CURVES OF RELUCTIVITY IN RELATION TO MAGNETIZING FORCE.

per gauss, and this increase, plotted graphically, would be represented by a straight line.

59. The accompanying curve sheet represents the results of actual observations by different observers upon different samples of soft wrought iron and cast iron. It will be seen that in the early stages of magnetization, below a critical magnetizing force, which varies with different samples from 1 to, perhaps, 15 gilberts-per-centimetre, corresponding to a prime magnetic intensity of 1 to 15 gausses (the latter in the case of cast iron), the reluctivity decreases with an increase in magnetizing force; but, when the critical magnetizing force is reached, the direction of the curve changes and the value becomes linear. Strictly speaking, the linear relation of reluctivity and magnetizing force, represented in the figure, is true only for the apparent reluctivity of the metal itself, and is irrespective of the ether which pervades the metal; for, were this relation strictly linear for all values of the magnetizing force beyond the critical value, the reluctivity would become infinite with an infinite magnetizing force; whereas, by observation, the reluctivity of the most highly saturated iron never exceeds unity, that of the air pump vacuum, or practically that of air. In point of fact we may consider the magnetism as being conducted through two paths in multiple; namely, that of the magnetic metal proper, and that of the ether permeating the metal. The first path may be called the *ferric path of metallic reluctivity*, and has a reluctance varying from a minimum at the critical magnetizing force, up to infinity, by the linear relation. The second is *the ether path of reluctivity*, and may be assumed to have a constant reluctivity of unity. The *joint reluctivity* of the two paths will be $\frac{1 \times \nu}{1 + \nu} = \frac{\nu}{1 + \nu}$ where ν, is the reluctivity of the ferric path. Since in actual dynamo machinery the value of the magnetizing force is never much more than 80 gilberts-per-centimetre, the above consideration is of small practical importance, since ν is, always much less than unity, say 0.01, and the discrepancy introduced by taking account of the multiple-connected ether path, is only the difference between 0.01 and $\frac{0.01}{1 + 0.01} = \frac{0.01}{1.01}$ or about 1 per cent., so that, for all practical purposes, we may assume that the metallic reluctivity is the actual reluctivity of the iron.

Beyond the critical magnetizing force, therefore, the value of the metallic reluctivity may be readily obtained by the equation $\nu = a + b \mathcal{H}$, where a, is the reluctivity which would exist at zero magnetizing force, if the linear relation held true below the critical value, and b, is the increase in reluctivity per gauss of prime magnetizing intensity expressed by $\mathcal{H}$. According to the present accepted values of the C. G. S. system, reluctivity is a numeric, and its value never exceeds unity; thus for wrought iron $a = 0.0004$, and $b = 0.000,057$.

60. If the ring shown in Fig. 44 be composed of wood, and be excited by 1,000 ampere-turns = 1,257 gilberts, then, since its mean length of circuit (circumference) is 60 cms., and cross sectional area 10 sq. cms., its reluctance will be 6 oersteds, the flux $\frac{1,257}{6} = 209.5$ webers, and the intensity $\frac{209.5}{10} = 20.95$ gausses, so that the magnetic force has a rate of descent of magnetic potential, the uniform distribution of which is $\frac{1,257}{60} = 20.95$ gilberts-per-centimetre. Strictly speaking, the intensity of the magnetic flux is not uniform over all portions of the area of cross section of the core, being denser at the inner circumference and weaker at the outer circumference. For example, if the inner circumference, instead of being 60 cms., which is the mean value, be 58 cms., the gradient of magnetic potential will be uniformly $\frac{1,257}{58} = 21.67$ gilberts-per-centimetre, and the intensity, 21.67 gausses; while, if the outer circumference be 62 cms., the intensity at that circumference will be $\frac{1,257}{62} = 20.27$ gausses. Since, however, all such existing differences of intensity can be made negligibly small, by sufficiently increasing the ratio of the size of the ring to its cross section, we may, for practical purposes, omit them from consideration.

61. Suppose now the core of the ring be composed of soft Norway iron instead of wood; then from the preceding curves, or the equation,

$$\nu = 0.0004 + 0.000,057\ \mathcal{H},$$

we find that at this mean intensity of $\mathcal{H} = 20.95$

$$\nu = 0.0004 + 0.001194 = 0.001594,$$

or about $\frac{1}{600}$th of that of air. The mean length of the circuit being 60 cms., and its area, as before mentioned, 10 sq. cms., its reluctance, under these circumstances, will be $\frac{60}{10} \times 0.001594 = 0.009564$ oersted, and the flux in the circuit $\frac{1,257}{0.009564} = 131,430$ webers, with an intensity of $\frac{131,430}{10} = 13,143$ gausses.

62. If the core of the ring instead of being of soft Norway iron be made of cast iron, the reluctivity, at $\mathcal{H} = 20.95$, would be approximately, 0.0046, and the reluctance of the circuit 0.0276 oersted, making the total flux 45,540 webers, with an intensity of 4,554 gausses, or about three times less than with soft Norway iron. The practical advantages, therefore, of constructing cores of soft Norway iron, rather than of cast iron, is manifest, when a high intensity is required.

63. It is important to remember that the entire conception of metallic reluctivity is artificial, and that although very convenient for purposes of computation, yet as already pointed out, it is incompetent to deal with the case of residual magnetism. Thus, if the prime M. M. F. from an iron ring be withdrawn, we should expect the flux to entirely disappear, whereas we know that a large proportion will generally remain. Since, however, electro-dynamic machinery rarely calls residual magnetism into account, the reluctivity theory is adequate for practical purposes beyond critical magnetizing forces.

64. As another illustration, let us consider a very long rod of iron, wound with a uniform helix. Here, as we have already seen, disregarding small portions near the extremities, the intensity may be regarded as uniform within the helix. Since the reluctance of the external circuit may be neglected, this flux density is $1.257\ n\ i$, gausses, where n, is the number of loops in the helix per centimetre of length, and i, is the exciting current strength in amperes. Or, regarding the intensity as being numerically equal to the gradient of magnetic potential, which changes steadily by $1.257\ n\ i$, per centimetre (this being the number of gilberts added in the circuit per centimetre of length, the fall of potential or drop in the external circuit being negligible), the gradient, within the helix, is $1.257\, n\, i$ gausses as before. A rod of Norway iron 1 metre long and 2 cms. in diameter, wound with twenty turns of wire to the centimetre, carrying a current of 1 ampere, would, at this magnetizing force, have an intensity in it of approximately $1.257 \times 20 \times 1 = 25.14$ gausses. The reluctivity of Norway iron would be by the preceding formula

$$\nu = 0.0004 + 0.000{,}057 \times 25.14 = 0.001833 \text{ or about } \frac{1}{500}\text{th}$$

of air. The length of rod being 100 cms., and its cross section 3.1416 square cms., the reluctance would be approximately $\frac{100}{3.1416} \times 0.001833 = 0.05836$ oersted. The total M. M. F. being $100 \times 20 \times 1 = 2{,}000$ ampere-turns $= 2{,}514$ gilberts. The flux in the circuit, assuming that the reluctance of the air path outside the bar may be neglected, is, approximately, $\frac{2{,}514}{0.05836} = 43{,}070$ webers, with an intensity of $\frac{43{,}070}{3.1416} = 13{,}710$ gausses.

65. In cases where the flux is confined to definite paths, as in a closed circular coil, or in a very long, straight, and uniformly wrapped bar, the preceding calculations are strictly applicable. When, however, an air-gap is introduced into the closed ring, that is, when its circuit becomes aero-ferric, the results begin to be vitiated, partly owing to the influence of diffusion, and partly to the results of the C. M. M. F. which is established at the air-gap. As the length of the air-gap increases, the degree of accuracy which can be attained by the application of the formula diminishes, but in dynamos, the aero-ferric circuits are in almost all cases of such a character, that the degree of approximation, which can be reached by these computations, is sufficient for all practical purposes; for, while it is impossible strictly to compute the magnetic circuit of a dynamo by any means at present within our reach, yet the E. M. F. of dynamos, and the speed of motors, can be predicted by computation within the limits of commercial requirements.

66. If the ring of Fig. 45 be provided with a small air-gap of 0.5 cm. in width, the intensity in the circuit, before the introduction of the iron core, will be practically unchanged by the existence of the gap, that is to say, with the same 1,000 ampere-turns, or 1,257 gilberts of M. M. F., the prime intensity existing in the ring will be practically 20.95 gausses. In the air-gap itself, the intensity will be less than this, owing to lateral diffusion of the flux; but, neglecting these influences, we may consider the intensity to be uniform. Now, introducing a soft, Norway iron core into the ring, the iron is subjected to an intensity of approximately, 20.95 gausses throughout the circuit. The reluctivity of the iron at this intensity, is, as we have seen, 0.001596. The length of the circuit in the iron will be 59.5 cms., and its cross section 10 sq. cms., making the ferric reluctance $\frac{59.5}{10} \times 0.001594 = 0.009484$ oersted. The reluctance of the air-gap, neglecting the influence of lateral diffusion, will be $\frac{0.5}{10} \times 1 = 0.05$ oersted, and the total reluctance of the circuit therefore, will be $0.009484 + 0.05 = 0.059484$ oersted. The flux in the circuit will be $\frac{1,257}{0.059484} = 21,130$ webers, and the intensity in the iron, 2,113 gausses. The existence of the air-gap has, therefore, reduced the flux from 131 kilowebers to 21 kilowebers.

67. In practical cases, however, the problem which presents itself is not to determine the amount of flux produced in a magnetic circuit under a given magnetizing force, but rather to ascertain the M. M. F., which must be impressed on a circuit in order to obtain a given magnetic flux. When the total required flux in a circuit is assigned, the mean intensity of flux in all portions of the circuit is approximately determinable, being simply the flux divided by the cross section of the circuit from point to point. What is required, is the reluctivity of iron at an assigned flux density and this we now proceed to determine.

From the equations, $\nu = a + b\mathcal{H}$, and $\mathcal{B} = \frac{\mathcal{H}}{\nu}$, corresponding in a magnetic circuit, to $i = \frac{e}{\rho}$ in the electric circuit, i, being the electric flux density or amperes-per-sq.-cm. and ρ, the resistivity, we obtain, $\nu = \frac{a}{1 - b\mathcal{B}}$.

This equation gives the reluctivity of any magnetic metal for any value of the flux density $\mathfrak{B}$ passing through it, when the value of the constants a and b, have been experimentally determined. The values of ν, so obtained are only true for reluctivities beyond the critical value, where the linear relation expressed in the equation $\nu = a + b\,\mathcal{H}$ commences.

68. The following table gives the values of the reluctivity constants a and b, for various samples of iron:

Sample.	a	b	Observer.
Soft Iron,	0.000,2	0.000,056	Stoletow.
Norway Iron,	0.000,1	0.000,059	Rowland.
Sheet Iron,	0.000,2375	0.000,0595	Fessenden.
" "	0.000,2275	0.000,0654	"
" "	0.000,3325	0.000,064	"
" "	0.000,213	0.000,05605	"
Cast Steel,	0.000,45	0.000,05125	"
" "	0.000,314	0.000,0563	"
Mitis Iron,	0.000,25	0.000,0575	"
Cast Iron,	0.001,031	0.000,129	"
Improved Cast Iron,	0.000,9025	0.000,106	"
Wrought Iron,	0.000,22	0.000,058	Hopkinson.
Dynamo Wrought Iron,	0.000,4	0.000,057	Kennelly.
" Cast Iron,	0.002,6	0.000,093	"
Annealed Norway Iron,	0.000,3	0.000,057	"

69. Fig. 47 shows curves of reluctivity of various samples of iron and steel at different flux densities. The descending branches are of practically little importance in connection with dynamo-electric machinery. They are included in the curves, however, in order to bring these into coincidence with actual observations. It will be seen, that while the reluctivity of Norway iron is only 0.000,5 at 8 kilogausses, that of cast iron is commonly about 0.010, or twenty times as great.

70. In order to show the application of the above curves of reluctivity, we will take the simplest case of the ferric circuit; namely, that of a soft Norway iron anchor ring, shaped as shown in Fig. 44, of 10 square centimetres cross section and 60 cms. mean circumference, uniformly wrapped with insulated wire. If it be required to produce a total flux of 80 kilowebers in this circuit, the intensity in the iron will be 8 kilogausses,

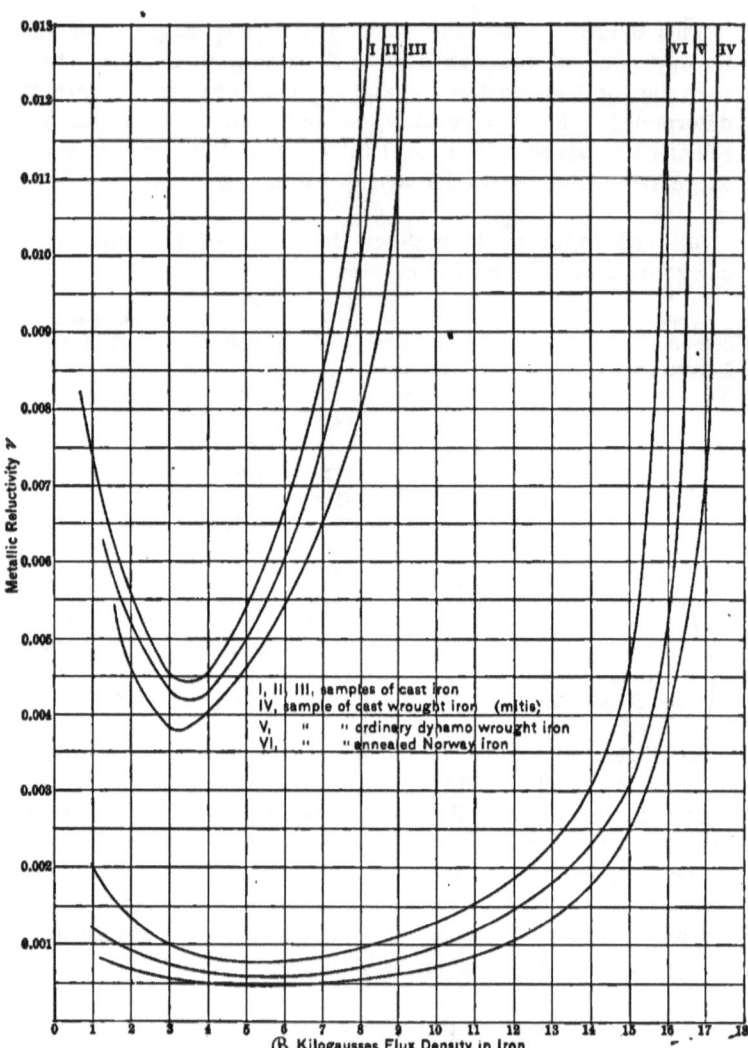

FIG. 47.—CURVES OF RELUCTIVITY IN RELATION TO FLUX DENSITY.

and, by following the curve for Norway iron, in Fig 47, it will be seen that its reluctivity at this density is 0.000,5. The reluctance of the circuit, therefore, will be $\frac{60}{10} \times 0.000,5 = 0.003$

oersted, and the M. M. F. necessary to produce the requisite magnetic flux will be $\mathfrak{F} = \Phi \mathfrak{R} = 80,000 \times 0.003 = 240$ gilberts, or $240 \times 0.7958 = 191$ ampere-turns.

71. If, however, the ring be of cast iron, instead of soft Norway iron, its reluctivity at this density would be say 0.010, and its reluctance $\frac{60}{10} \times 0.010 = 0.06$ oersted, from which the required M. M. F. will be $80,000 \times 0.06 = 4,800$ gilberts $= 3,820$ ampere-turns. The importance of employing soft iron for ferric magnetic circuits, in which a large total flux is required, will, therefore, be evident.

CHAPTER VI.

AERO-FERRIC MAGNETIC CIRCUITS.

72. We will now consider the case of the aero-ferric magnetic circuit. Fig. 48 is a representation of a simple ferric circuit consisting of two closely fitting iron cores, the upper of which is wrapped with a magnetizing coil M. The polar surfaces are made to correspond so closely, that when the coil M, has a magnetizing current sent through it, the magnetic attraction between the two cores will cause them to exclude all sensible air-gaps. The general direction of the flux-paths is shown by the dotted arrows, and a mechanical stress is exerted within the iron along the flux-paths.

These stresses cannot be rendered manifest, so long as the iron is continuous. In other words, the continuous anchor ring, as shown in Fig. 44, would give no evidence of the existence of stress along its flux-paths. In the case shown in Fig. 48, the stress is rendered evident by the force which must be applied to the two magnetized cores in order to separate them. The amount of this force depends upon the magnetic intensity in the iron at the polar surfaces, and, if $\mathcal{B}$, represents this intensity in gausses, the attractive force exerted along the flux-paths at the polar surfaces; *i. e.*, perpendicularly across them, will be $\frac{\mathcal{B}^2}{8\pi}$ dynes-per-square-centimetre of polar surface. The *dyne* is the *fundamental unit of force* employed in the system of C. G. S. units universally employed in the scientific world, and is equal to the weight of 1.0203 milligrammes at Washington; that is to say, the attractive force which the earth exerts upon one milligramme of matter, is approximately, equal to one dyne.

73. If the magnetic circuit shown in Fig. 48 has a uniform area of cross section of 12 square centimetres, and the magnetic intensity in the circuit be everywhere 17 kilogausses, then the attractive force exerted across each square centimetre of the polar surfaces at R_1 and R_2, will be

$$\frac{17{,}000 \times 17{,}000}{8 \times 3.1416} = 11{,}500{,}000 \text{ dynes,}$$

or $11{,}500{,}000 \times 1.0203 = 11{,}730{,}000$ milligrammes weight = 11,730 grammes weight = 25.86 lbs. weight.

As there are twelve square centimetres in each polar surface,

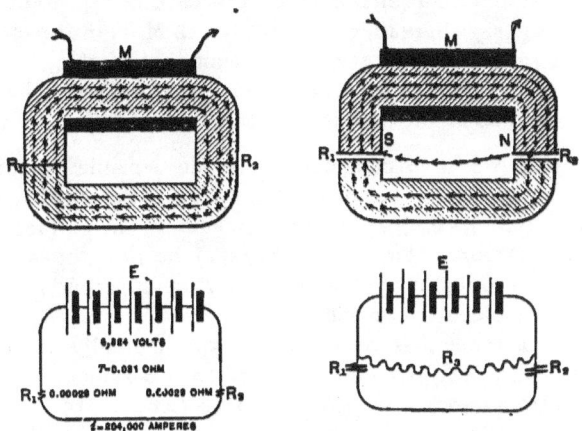

FIGS. 48 AND 49.—DIAGRAMS REPRESENTING A SIMPLE FERRIC, AND AN AERO-FERRIC, CIRCUIT, AND THEIR ELECTRIC ANALOGUES.

the total pull across each gap will be $12 \times 25.86 = 310.32$ lbs. weight; and since there are two gaps, the total pull between the iron cores will be 620.64 lbs. weight, so that, if the whole magnet were suspended in the position shown in Fig. 48, this weight should be required to be suspended from the lower core (less, of course, the weight of the lower core) in order to effect a separation; or, in other words, this should be the maximum weight which the magnet could support.

74. In order to ascertain the M. M. F. needed to produce the required intensity of 17 kilogausses through the circuit in order to cause this attraction, we find, by reference to Fig. 47, that the reluctivity of Norway iron at this intensity is 0.0073; *i. e.*, $\frac{73}{10{,}000}$ that of air. The reluctance of the magnetic cir-

cuit will, therefore, be $\frac{50}{12} \times 0.0073 = 0.03042$ oersted. The total flux through the circuit will be $17,000 \times 12 = 204,000$ webers, and the M. M. F. required to produce the flux, therefore, will be $204,000 \times 0.03042 = 6,206$ gilberts, or $6,206 \times 0.7958 = 4,937$ ampere-turns. If, then, the coil M, has 2,000 turns, it will be necessary to send through it a current of 2.469 amperes, in order to produce the flux required.

The electric circuit analogue of this case is represented in the same figure, where E, represents the E. M. F. in the electric circuit as a voltaic battery, and the amount of this E. M. F. necessary to produce a current of strength i, amperes, when the total resistance of the circuit is r, ohms, will be $E = i\,r$ volts.

75. So far we have considered that no sensible reluctance existed at the polar surfaces R_1, and R_2. Practically, however, it is found, that, no matter how smooth the surfaces may be, and, therefore, how closely they may be brought into contact, a small reluctance does exist, owing, apparently, to the absence of molecular continuity.

This reluctance has been found experimentally, in case of very smooth joints, to be equivalent to the reluctance of an air-gap, from 0.003 to 0.004 cm. wide (0.0012" to 0.0016"). Taking this reluctance into account we have at R_1, and at R_2, an equivalent reluctance of air path, say 0.0035 cm. long and 12 cms. in cross-sectional area. Since the reluctivity of air is unity, the reluctance at each gap becomes $\frac{0.0035}{12} \times 1 =$ 0.000,29 oersted, and the reluctance of the circuit has, therefore, to be increased by 0.000,58 oersted, making a total of $0.03042 + 0.000,58 = 0.031$ oersted, and requiring the M. M. F. of $204,000 \times 0.031$ or $6,324$ gilberts $= 5,032$ ampere-turns, or an increase of current strength to 2.516 amperes.

76. It is evident, since the attractive force exerted across a square centimetre of polar surface is equal to $\frac{\mathcal{B}^2}{8\pi}$ dynes, that doubling the intensity at the polar surface will quadruple the attraction per square centimetre. Therefore, all electro-magnets, which are intended to attract or support heavy weights, are designed to have as great a cross-sectional area of polar surface as possible, combined with a high magnetic intensity across these surfaces. If, however, the increase of the area of polar surface is attended by a corresponding diminution of flux density, the total attractive force across the surface will be diminished, because the intensity, per-unit-area, will be reduced in the ratio of the square of the intensity, while the pull will only increase directly with the surface. It is evident, therefore, that soft iron of low reluctivity is especially desirable in powerful electro-magnets.

If, for example, cast iron was employed in the construction of the magnet of Fig. 48, instead of soft Norway iron, and the same M. M. F., namely, 6,324 gilberts were applied, the mean magnetizing force would be this M. M. F., divided by the mean length of the circuit in cms., or $\mathcal{K} = \dfrac{6,324}{50} = 126.48$ gilberts-per-centimetre.

At this magnetizing force, a sample of cast iron would have a reluctivity represented by the formula $\nu = (a + b\mathcal{K})$, where a, may be 0.0027, and b, 0.000,09, so that its reluctivity at 133.92 gilberts per centimetre of magnetizing force would be (0.0027 + 0.000,09 × 126.48) = 0.01407. The reluctance of the cast iron circuit, including the small reluctance in the air-gaps, would be $\dfrac{50}{12}$ × 0.01407 = 0.05863 oersted, and the flux in the circuit would be $\dfrac{6,374}{0.05863}$ = 108,700 webers, or an intensity of 9,058 gausses. The magnetic attraction between the surfaces per-square-centimetre, would, therefore, be $\dfrac{9,058 \times 9,058}{25.133}$ = 3,264,000 dynes, or 3,331 grammes weight, or 7.342 lbs. weight; and, since the total polar surface amounts to 24 square centimetres, the total attractive force exerted between and across them is 176.2 lbs. weight. The effect of introducing cast iron instead of wrought iron into the magnetic circuit, keeping the dimensions and M. M. F. the same, has, then, been to reduce the total pull from 620.64 lbs. to 176.2 lbs., or 71.6 per cent.

77. If now an air-gap be placed in the circuit at R_1, and R_2, of half an inch (1.27 cm.) in width, as in Fig 49, two results will follow; viz.,

(1.) A greater reluctance will be produced in the circuit.

(2.) A leakage or shunt path will now be formed through the air between the poles N and S. Strictly speaking, there will be some leakage in the preceding case of Fig. 48, but with a ferric circuit of comparatively short length, it will have been so small as to be practically negligible. In Fig. 49, however, the reluctance of the main circuit between the poles including the air-gaps will be so great as to give rise to a considerable difference of magnetic potential between the poles N and S, so that appreciable leakage will occur between these points. The reluctance of the leakage-paths through the air will usually be very complex, and difficult to compute, but, in simple geometrical cases, it may be approximately obtained without great difficulty. In this case we may proceed to determine the magnetic circuit first on the assumption that no leakage exists, and second on the assumption of the existence of a known amount of leakage. Assuming that the cores are of soft Norway iron, and that it is required to establish a total flux of 204,000 webers through the circuit, then the flux density in the iron will be 17 kilogausses and its reluctivity 0.0073. The reluctance of the circuit, so far as it is composed of iron, will be 0.03042 oersted, while the reluctance of each air-gap will be $\frac{1.27}{12} \times 1 = 0.1058$; or, in all, 0.2016 oersted. The total reluctance of the circuit will, therefore, be 0.23202 oersted, and the M. M. F. required will be $204,000 \times 0.23202 = 47,330$ gilberts $= 37,660$ ampere-turns; or, with 2,000 turns, 18.83 amperes. The attractive force on the armature will be 620 lbs. as in the previous case.

78. Considering now the effect of leakage, we may assume that the reluctance of the leakage path through the air R_s, is 0.5 oersted, and that a flux of 108 kilowebers has to be produced through the lower core; the length of mean path in the lower core being 20 cms., and in the upper core 30 cms., it is required to find the M. M. F., which will produce this flux through the lower core.

The intensity in the lower core will be $\frac{108,000}{12} = 9,000$ gausses, at which intensity the reluctivity of Norway iron will be, by Fig. 47, 0.000,6, so that the reluctance of the lower core will be $\frac{20}{12} \times 0.000,6 = 0.001$ oersted, and this added to the reluctance of the two air-gaps, 1.27 cms. in width, $= 0.2016 + 0.001 = 0.2026$ oersted. The magnetic difference of potential in this branch of the double circuit will, therefore, be $108,000 \times 0.2026 = 21,880$ gilberts. This will also be the difference of magnetic potential between the terminals of the leakage path R_y, and the leakage flux will, therefore, be $\frac{21,880}{0.5} = 43,760$ webers. The total flux in the main circuit through the upper core will be the sum of the flux in the two branches, or $108,000 + 43,760 = 151,760$ webers, making the intensity in the upper core $\frac{151,760}{12} = 12,647$ gausses, at which intensity the reluctivity is 0.00121, so that the reluctance of the upper core is $\frac{30}{12} \times 0.0012 = 0.003$ oersted. The drop of potential in the upper core will, therefore, be $151,760 \times 0.003 = 455$ gilberts, and the total difference of potential in the circuit, or the M. M. F., will be $21,880 + 455 = 22,335$ gilberts $= 17,775$ ampere-turns, or 8.89 amperes at 2,000 turns.

79. It is obvious that the results obtained by the preceding method of calculation cannot be strictly accurate, since no account has been taken of any magnetic leakage except that which occurs directly between the poles N and S. Also we have assumed that the flux density remains uniform throughout the lengths of the two cores. When a greater degree of accuracy is desired, corrections may be introduced for the effects of these erroneous assumptions, but the examples illustrate the general methods by which the magnetic circuits of practical dynamo-electric machines may be computed with fair limits of accuracy.

CHAPTER VII.

LAWS OF ELECTRO-DYNAMIC INDUCTION.

80. When a conducting wire is moved through a magnetic flux, there will always be an E. M. F. induced in the wire, unless the motion of the wire coincides with the direction of the flux; or, in other words, unless the wire in its motion does

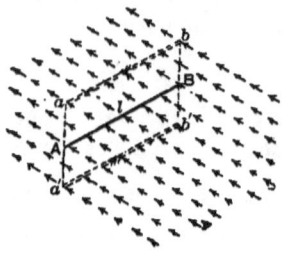

FIG. 50.—CONDUCTOR PERPENDICULAR TO UNIFORM MAGNETIC FLUX, AND MOVING AT RIGHT ANGLES TO SAME.

not pass through or cut the flux. Thus, if, as in Fig. 50, a straight wire $A\ B$, of l cms. length, extending across a uniform flux, be moved at right angles to the flux, either upwards or downwards, to the position, for example, $a\ b$, or $a'\ b'$, it will have an E. M. F. induced in it, the direction of which will change with the direction of the motion.

81. A convenient rule for memorizing the direction of the E. M. F. induced in a wire cutting, or moving across, magnetic flux, is known as *Fleming's hand rule.* Here, as in Fig. 51, the right hand being held, with the thumb, the forefinger and the middle finger extended as shown, the thu*m*b being so pointed as to indicate the direction of *m*otion, and the *f*orefinger the direction of the magnetic *f*lux, then the *m*iddle finger will indicate the direction of induced E. M. F. For example, if, as in

Fig. 50, a wire be moved vertically downwards from *A B*, to *a' b'*, and the thumb be held in that direction, the forefinger pointing in the direction of the flux, the E. M. F. induced in the wire will take the direction *a' b'*, during the motion, following the direction of the middle finger. If, however, the wire be moved upwards through the flux, an application of the same

FIG. 51.—FLEMING'S HAND RULE.

rule will show that the direction of the induced E. M. F., as indicated by the middle finger, is now changed.

82. The induction of electromotive force in a conductor, moving so as to pass through or cut magnetic flux, is called *electro-dynamic induction*. The value of the E. M. F. induced in a wire by electro-dynamic induction depends,

(1.) On the density of the magnetic flux.
(2.) On the velocity of the motion, and
(3.) On the length of the wire.

This is equivalent to the statement that the E. M. F., induced in a given length of wire, depends upon the total amount of flux cut by the wire per second in the same direction; or,

$$e = \mathfrak{B}\,l\,v \qquad \text{C. G. S. units of E. M. F.}$$

Where $\mathfrak{B}$, is the intensity of the flux in gausses, l, the length of the conductor in cms., v, the velocity of motion in cms.-per-second, and e, the induced electromotive force as measured in C. G. S. units. Since one international volt is equal to

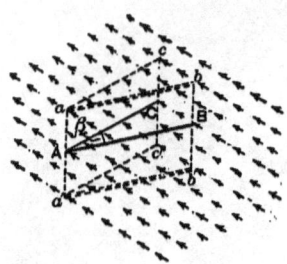

FIG. 52.—CONDUCTOR OBLIQUE TO UNIFORM MAGNETIC FLUX, AND MOVING AT RIGHT ANGLES TO SAME.

100,000,000 C. G. S. units of E. M. F., the E. M. F. induced in the wire will be

$$e = \frac{\mathfrak{B}\,l\,v}{100{,}000{,}000} \qquad \text{volts.}$$

83. The preceding equation assumes that the wire is not only lying at right angles to the flux, but also that it is moved in a direction at right angles to the direction of the flux. If instead of being at right angles to the flux, the wire makes an angle β, with the perpendicular to the same, as shown in Fig. 52, then the length of the wire has to be considered as the virtual length across the flux, or as its projection on the normal plane, so that the formula becomes,

$$e = \frac{\mathfrak{B}\,l\,v\cos\beta}{100{,}000{,}000} \qquad \text{volts.}$$

If the motion of the wire, instead of being directed perpendicularly to the flux, is such as to make an angle α, with the perpendicular plane, the effective velocity is that virtually taking place perpendicular to the flux, or $v \cos \alpha$, as shown in Fig. 53, so that the formula becomes in the most general case,

$$e = \frac{\mathcal{B} \, l \cos \beta \, v \cos \alpha}{100,000,000} \quad \text{volts}$$

84. It will be seen that in all cases the amount of flux cut through uniformly in one second, gives the value of the E. M. F.

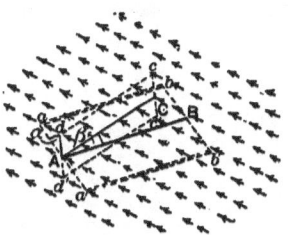

FIG. 53.—CONDUCTOR OBLIQUE TO UNIFORM MAGNETIC FLUX, AND MOVING OBLIQUELY TO SAME.

induced in the wire, and that the value of the E. M. F. does not depend upon the amount of flux that has been cut through, or that has to be cut through, but upon the instantaneous rate of cutting. The E. M. F. ceases the moment the cutting ceases.

85. If the loop $A \, B \, C \, D$, Fig. 54, be rotated about its axis $O \, O'$, in the direction of the curved arrows, then, while the side $C \, D$, is ascending, the side $A \, B$, is descending; consequently, the E. M. F. in the side $C \, D$, will be oppositely directed to the E. M. F. in the side $A \, B$. Applying Fleming's hand rule to this case, we observe that the directions of these E. M. Fs. are as indicated by the double-headed arrows, and, regarding the conductors $C \, D$ and $A \, B$, as forming parts of the complete circuit $C \, D \, A \, B$, it is evident that the E. M. Fs. induced in $A \, B$ and $C \, D$, will aid each other, while, if they are permitted to produce a current, the current will flow through the circuit in the same direction.

86. We have seen that no E. M. F. is induced in a wire unless it cuts flux. Consequently, the portions $B \, C$ and $A \, D$, of the circuit which move in the plane of the flux, will contribute nothing to the E. M. F. of the circuit.

If the dimensions of the wires forming this loop shown in the figure, are such that CD and AB, having each a length of 12 cms., while AB and BC, are 4 cms. each., the circumference traced by the wires AB and CD, in their revolution about the axis, will be $3.1416 \times 4 = 12.567$ cms.; and, if the rate of rotation be 50 revolutions per second, the speed with which the wires AB and CD, revolve will be 628.3 cms. per second. If the intensity of the magnetic flux B, is uniformly 5 kilogausses, the E. M. F. induced in each of the wires AB

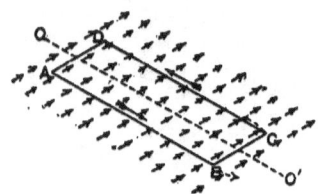

FIG. 54.—RECTANGULAR CONDUCTING LOOP ROTATING IN UNIFORM MAGNETIC FLUX.

and CD, will be, $5,000 \times 12 \times 628.32 = 37,699,200$ C. G. S. units of E. M. F., or 0.377 volt. This value of the E. M. F. only exists at the instant when the loop has its plane coincident with the plane of the flux, and the sides cut the flux at right angles. In any other position, the motion of these sides is not at right angles to the flux, so that the E. M. F. is reduced.

87. In order that the E. M. F. induced in a wire may establish a current in it, it is necessary that such wire should form a complete curcuit or loop, as indicated in Fig. 55. When such a conducting loop is moved in a magnetic field, some or all portions of the loop will cut flux, and will thereby contribute a certain E. M. F. around the loop. If the loop moves in its own plane, in a uniform magnetic flux, there will be no resultant E. M. F. generated in it. For example, considering a circular loop, we may compare any two diametrically opposite segments, when it is evident that each member of such a pair cuts through the same amount of flux per second, and will, therefore, generate the same amount of E. M. F., but in directions opposite to each other in the loop. At the same time, it is clear that

the total amount of flux in the loop does not change; for, while the flux is being left by the loop at its receding edge, it is entering the loop at the same rate at its advancing edge, and, since these two quantities of flux are equal, the total amount of flux enclosed by the loop remains constant.

88. The cutting of flux by the edges of a moving loop, therefore, resolves itself into the more general condition of enclosing flux in a loop. The value of the E. M. F. induced around

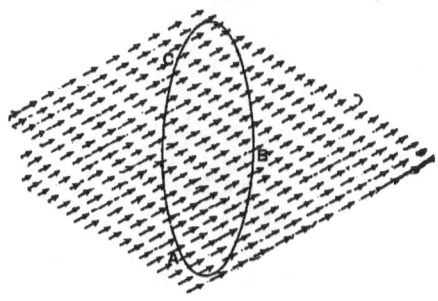

FIG. 55.—CIRCULAR CONDUCTING LOOP PERPENDICULAR TO UNIFORM MAGNETIC FLUX.

the loop does not depend upon the actual quantity of flux enclosed, but on the rate at which the enclosure is being made. If, as we have already seen, the loop is so moved that the total flux it encloses undergoes no variation, the amount entering the loop being balanced by the amount leaving it, although E. M. Fs. will be induced in those parts of the loop where the flux is entering and where it is leaving, yet these E. M. Fs. being opposite, exactly neutralize each other, and leave no resultant E. M. F. Consequently, the value of the E. M. F. induced at any moment in the loop by any motion, does not depend upon the flux density within the loop, but on the rate of change of flux enclosed.

89. If Φ, be the total flux in webers contained within a single loop, such as shown at $A\ B\ C$, in Fig. 55, the mean rate at which this flux is changing during any given period of time, will be the quotient of the change in the enclosure, divided by that amount of time, so that if Φ, changes by 20,000 webers in two seconds, the mean rate of change during that time will be 10,000 webers per second, and this will be the E. M. F. in the loop expressed in C. G. S. units. But, during these two seconds of time, the change may not have been progressing uniformly, in which case only the average E. M. F. can be stated as being equal to the 10,000 C. G. S. units. Where the change is not uniform, the rate at any moment has to be determined by taking an extremely short interval, so that if dt, represents

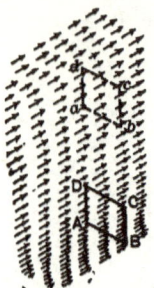

FIG. 55A.—RECTANGULAR CONDUCTING LOOP IN NON-UNIFORM MAGNETIC FLUX.

this indefinitely small interval of time, and $d\Phi$, the corresponding change in the flux enclosed during that interval in webers, the rate of change will be $\dfrac{d\Phi}{dt}$ webers-per-second, and this will be the value of the induced E. M. F. at each instant.

90. If a small square loop of wire $A\,B\,C\,D$, one cm. in length of edge, placed at right angles to the flux as shown in Fig. 55A, contains a total quantity of flux amounting to 10,000 webers, the mean flux density at the position occupied by the square, will be 10,000 gausses. If now, the loop be moved uniformly upward in its own plane to the position $a\,b\,c\,d$, so as to accomplish the journey in the $\dfrac{1}{100}$ th part of a second, and if the flux enclosed by the loop at the position $a\,b\,c\,d$, be 1,000 webers, then 9,000 webers will have escaped from the loop during the motion. Assuming that the distribution of flux density in the field was such that the emission took place uniformly, the E. M. F. in the loop, during the passage, will have been,

$$\frac{\Delta \Phi}{\Delta t} = \frac{9,000}{\frac{1}{100}} = 900,000 \text{ C. G. S. units} = 0.009 \text{ volt.}$$

91. If, however, the rate of emptying, during the motion, were not uniform, 0.009 volt would be the average E. M. F., and not the E. M. F. sustained during the interval; or, in other words, the instantaneous value of the E. M. F. in the loop would vary at different portions of this short interval of time, or at corresponding different positions during the journey; but, in all cases, the time integral of the E. M. F. will be equal to the change in Φ; thus, the change in Φ, is, in this case, 9,000 webers. If the motion is made in $\frac{1}{100}$th of a second, the E. M. F., will be 900,000 C. G. S. units of E. M. F., which, multiplied by the time (0.01 second), gives 9,000 webers. If, however, the motion were uniformly made in half a second, the E. M. F. would have been 18,000 C. G. S. units, which, multiplied by the time, would give as before 9,000 webers; and under whatever circumstances of velocity the change were made, the sum of the products of the instantaneous values of E. M. F. multiplied into the intervals of time during which they existed, would give the total change in flux of 9,000 webers. Or in symbols,

$$\text{Since } e = \frac{d\Phi}{dt}$$

$$\int e\, dt = \Delta \Phi$$

The first equation simply expresses that the E. M. F., e, is the instantaneous rate of change in the flux enclosed, and the second equation shows that the difference in the enclosure between any two conditions of the loop is the time integral of the E. M. F., which has been induced in the loop during the change, assuming of course, that the change continues in the same direction; *i. e.*, that the flux through the loop has continually increased or decreased.

92. If a circuit contains more than one loop, as, for example, when composed in whole, or in part, of a coil, the turns of which are all in series, the E. M. F. induced in any one turn or loop of the coil, may be regarded as being established independently of all the other loops, so that the total E. M. F. in the circuit will be the sum of all the separate E. M. Fs. existing at any instant in the loops, and may, therefore, be regarded as the instantaneous rate of change in the flux linked with the entire circuit. A coil, therefore, may be regarded as a device for increasing the amount of flux magnetically linked with an electric circuit, so that by increasing the number of loops of conductor in the circuit, the value of the induced E. M. F. corresponding to any change in the flux, is proportionally increased, and if the coil or system of loops forming the cir-

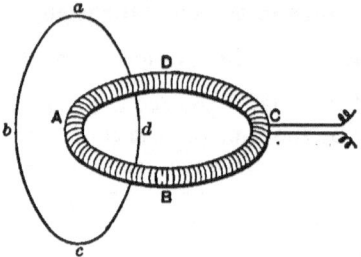

FIG. 56.—CLOSED CIRCULAR HELIX LINKED WITH A LOOP OF WIRE.

cuit, contains in the aggregate Φ webers of flux linked with it, taking each turn separately and summing the enclosures, then the time integral of E. M. F. in the circuit will be the total change in Φ, and this will be true, whether the loop is changing its position, or whether the flux is changing in intensity or in direction.

93. It is evident from the preceding, that there are two different standpoints from which we may regard the production of electromotive force in a conducting circuit by electrodynamic induction; namely, that of cutting magnetic flux, and that of enclosing magnetic flux. These two conceptions are equivalent, being but different ways of regarding the same phenomenon. The amount of flux enclosed by a loop can only vary by the flux being cut at the entering edge or edges at a different rate to that at the receding edge; or, in mathematical language, the surface integral of enclosing is equal to the line integral of cutting, taken once round the loop. This statement is equally true whether the flux is at rest and the conductor moving, or the conductor at rest and the flux moving, or whether both conductor and flux are in relative motion.

94. Cases of electro-dynamic induction may occur where the equivalence of cutting and enclosing magnetic flux apparently fails. On closer examination, however, the equivalence will be manifest. For example, in Fig. 56, let *A B C D* be a wooden anchor ring uniformly wound with wire, as shown in Fig. 44, and *a b c d*, a circular loop of conductor linked with the ring.

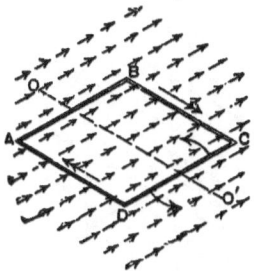

FIG. 57.—SQUARE CONDUCTING LOOP ROTATED IN UNIFORM FLUX. FIRST POSITION.

It has been experimentally observed that when a powerful current is sent through the winding of the anchor ring, no appreciable magnetic flux is to be found at any point outside the ring, although within the core of the ring a powerful magnetic flux is developed. Nevertheless, both at the moment of applying and at the moment of removing the exciting current through the winding of the ring, an E. M. F. is induced in the loop *a b c d*, whose time integral in C. G. S. units, is the total number of webers of change of flux in the ring core. It might appear at first sight that this E. M. F. so induced in the loop cannot be due to the cutting of flux by the loop, but must be due to simple threading or enclosing of flux. It is clear, however, that the mere act of enclosure will not account for the induction of the E. M. F., since the passage of flux through the centre of the loop cannot produce E. M. F. in the loop itself, unless activity is transmitted from the centre of the loop to its periphery. In other words, action at a distance, without intervening mechanism of propagation, is believed to be impossible.

Could we see the action which occurs when the current first passes through the ring-winding, we should observe flux apparently issuing from all parts of the ring and passing into surrounding space, at a definite speed. The loop *a b c d*, would receive the impact of flux from the adjacent portions of the ring before receiving that from the more distant parts of the ring, and, in this sense, would actually be cut by the flux. As soon as the flux has become established, and the current in

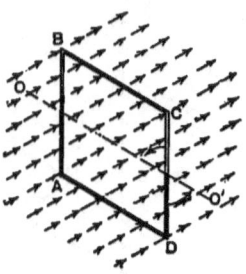

FIG. 58.—SQUARE CONDUCTING LOOP ROTATED IN UNIFORM FLUX. SECOND POSITION.

the winding steady, it is found that the flux from any particular portion of the ring is equal and opposite to that from the remainder of the ring, and is, therefore, cancelled or annulled at all points except within the ring core. It is evident, therefore, that we may regard the E. M. F induced in the loop *a b c d* as due either to the cutting of the boundary by flux, or to the enclosure of flux.

95. Let us consider the case of a square conducting loop *A B C D*, Fig. 57, having its plane parallel with the uniform magnetic flux shown by the dotted arrows. If this loop be rotated about the axis *O O'*, which is at right angles to the magnetic flux, and symmetrically placed with regard to the loop, so that *A D*, descends, and *B C*, ascends, these sides, which cut flux during the rotation, will have E. M. Fs. generated in them, in accordance with Fleming's hand rule already described in Par. 81, and in the direction shown by the double arrows. The sides AB and DC, which do not cut flux during the motion, will add nothing to the E. M. F. generated. The figure shows that while the sides AD and CB, have oppositely directed E. M. Fs., yet regarding the entire loop as a conducting circuit, these E. M. Fs. tend to produce a current which circulates in the same direction.

96. As already pointed out, the value of the E. M. F. generated in the sides AD and CB, of the loop, by the cutting of the flux, will depend upon the rate of filling and emptying the

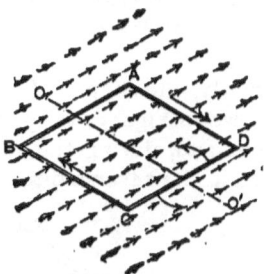

FIG. 59.—SQUARE CONDUCTING LOOP ROTATED IN UNIFORM FLUX. THIRD POSITION.

loop with flux, and it is evident that this rate is at a maximum when the loop is empty; *i. e.*, in the position it occupies in Fig. 57, when the plane of the loop coincides with the direction of the flux, and the motion of its sides is at right angles thereto; for, when the loop reaches the position shown in Fig. 58, namely, when it is full of flux; or, when its plane is as right angles to the flux, then at that instant the rotation of the loop neither adds to nor diminishes, the amount of flux enclosed, so that the E. M. F. in the loop is zero.

97. Continuing the rotation of the loop in the same direction, the E. M. F. generated will increase from this position until the position shown in Fig. 59 is reached, where the plane of the loop is again coincident with the plane of the flux, but in which the side AD, has moved through 180°, or one-half a revolution from the position shown in Fig. 57, and the directions of E. M. Fs. in the wire, as shown, will be changed so far as the wire is concerned, being now from A to D, instead of from D to A, in the conducting branch AD; and from C to B, instead of from B to C, in the conducting branch BC. The direction of E. M. F. around the loop, will, therefore, be

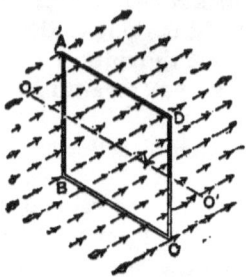

FIG. 60.—SQUARE CONDUCTING LOOP ROTATED IN UNIFORM FLUX. FOURTH POSITION.

reversed. Consequently, the loop $ABCD$, during its first half revolution as shown in Figs. 57 to 59, has an E. M. F. in it in the same direction; and, during the remaining half-revolution, has its E. M. F. in the reverse direction, as shown.

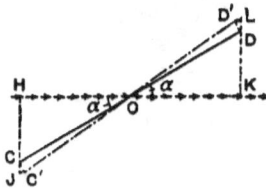

FIG. 61.—FLUX OBLIQUE TO PLANE OF ROTATING LOOP.

98. The value of the E. M. F. generated in a loop, during its rotation, depends upon the flux density, on the area of the loop, and on the rate of rotation.

Assuming the side of the loop CD, to occupy the position shown in Fig. 61, making an angle α, with the direction HK, of the flux, then the E. M. F. generated in the loop at this instant is the rate at which flux is being admitted into the loop. If l cms., be the length of the side of the loop or the length of AD, in Fig. 57, the amount of flux embraced at this instant will be $l \, \mathcal{B} \times 2\, DK$. During the next succeeding small interval of time dt, if the angular velocity of the loop, ω radians per second, carries it to the position $C'D'$, the amount of flux admitted during that time will be $l \, \mathfrak{B} \times 2\,D\,L$. But $D\,L = D\,D' \times$ cosine of angle $D'D\,L$, and this angle is equal to the

```
C                    D'
   ─────────────────
C'                    D
```

FIG. 62.—FLUX COINCIDENT WITH PLANE OF ROTATING LOOP.

angle α, so that $D\,L = D\,D' \times \cos \alpha$, and $D\,D'$, will be $\dfrac{l}{2}\omega\,dt$ cms. in length, since the radius $O\,D = \dfrac{l}{2}$; consequently, the flux admitted into the loop during this brief interval of time dt, will be

$$d\,\Phi = 2\,l \times \frac{l}{2}\,\mathfrak{B}\,\omega \cos \alpha\,dt,\ \text{or}\ l^2\,\mathfrak{B}\,\omega \cos \alpha\,dt$$
$$= \Phi\,\omega \cos \alpha\,dt$$

so that
$$\frac{d\,\Phi}{dt} = \Phi\,\omega \cos \alpha.$$

Thus, at the instant of time in which the loop has reached the

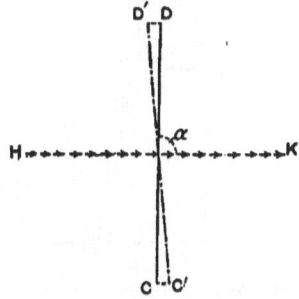

FIG. 63.—FLUX PERPENDICULAR TO PLANE OF ROTATING LOOP.

position $O\,D$, if α, be the angle which the loop makes at any time with the direction of the flux, the E. M. F. e, the instantaneous rate of increase in the flux, or will be generally expressed in C. G. S. units by

$$e = \Phi\,\omega \cos \alpha$$

Φ, being the maximum amount of flux in webers ($l^2\,\mathfrak{B}$), which the loop can embrace. When the plane of the loop coincides with the direction HK, of the flux, as shown in Fig. 62, DD', is brought into coincidence with DL, or the cosine of α is 1. So that the E. M. F. e, in the loop has a maximum value, and

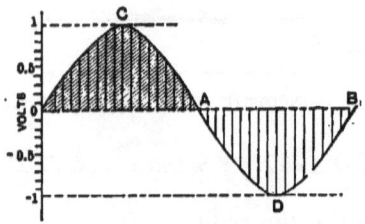

FIG. 64.—CURVE OF E. M. F. INDUCED IN ROTATING LOOP.

is equal to $\Phi \omega$, while when the loop is at right angles to the flux, or as shown in Figure 63, DD', the succeeding small

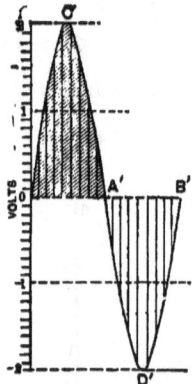

FIG. 65.—CURVE OF E. M. F. INDUCED IN LOOP ROTATING AT DOUBLED SPEED.

excursion of the loop, is at right angles to DL, or cosine $\alpha = 0$, so that $e = 0$.

99. If $\mathfrak{B}$, as in the case represented by Figs. 57 to 60, be two kilogausses, and $l = 100$ cms., then $\Phi = 100 \times 100 \times 2,000 = 20$

megawebers. If the loop be rotated in the direction shown at an angular velocity of 50 radians per second ($\frac{50}{2\pi}$ revolutions per second), the E. M. F. e., will be

$e = 20{,}000{,}000 \times 50 \times \cos \alpha$, or $100{,}000{,}000 \cos \alpha$
$ = 1 \cos \alpha$ volt.

The E. M. F. generated by the loop, therefore, varies periodically between 1, 0, — 1, 0, and 1. If these values be

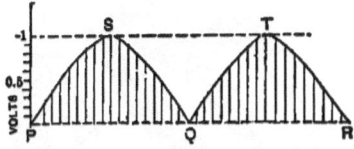

FIG. 66.—CURVE OF E. M. F. COMMUTED IN EXTERNAL CIRCUIT.

plotted graphically as ordinates, to a scale of time as abscissas, the curve shown in Fig. 64 will be obtained, where the distance $A\,O$, represents the time occupied by one half revolution of the loop, the E. M. F. being positive from O to A, and negative from A to B. If now, the speed of revolution be doubled; *i. e.*, increased to 100 radians per second, the time occupied in each revolution will be halved, and $O'A'$, Fig. 65, will be half the length of $O\,A$, but e, will be doubled as shown. The shaded area $O'\,C'\,A'$, in Fig. 65, is equal to the area $O\,C\,A$, of of Fig. 64. The E. M. F. generated by the loop is alternating, being positive and negative during successive half revolutions, but, by the aid of a suitable commutator, the E. M. F. can be made unidirectional in the external circuit, as represented in Fig. 66, where the curve $P\,S\,Q$, corresponds to $O\,C\,A$, in Fig. 64 and $Q\,T\,R$, to $A\,D\,B$.

CHAPTER VIII.

ELECTRO-DYNAMIC INDUCTION IN DYNAMO ARMATURES.

100. The type of curve represented in Figs. 64, 65, and 66, showing the E. M. F. generated by the rotation of a conducting loop in a uniform magnetic flux, may be produced by the rotation of the coil represented in Fig. 67. Here a number of circular loops, formed by winding a long insulated wire upon

FIG. 67.—COIL FOR INDUCING FEEBLE E. M. FS. BY REVOLUTION IN EARTH'S MAGNETIC FLUX.

a circular wooden frame, are capable of being rotated by the handle, in the uniform magnetic flux of the earth. If the mean area of the loops be 1,000 sq. cms., the number of loops 500, and the intensity of the earth's magnetic flux threading the loop 0.6 gauss, then the E. M. F. generated by rotating the loop will depend only on the speed of rotation. Assuming this to be 5 revolutions-per-second, or an angular velocity of $5 \times 2\pi = 15.708$ radians-per-second, the E. M. F. will vary between $+\Phi\omega$ and $-\Phi\omega$, in each half revolution. Here Φ, the total flux linked with the coil is $500 \times 1,000 \times 0.6 = 300,000$

webers, and $\omega = 15.708$, so that the maximum value of the E. M. F. generated in the coil will be 4,712,400 C. G. S units $= 0.047$ volt, or roughly $\frac{1}{20}$th volt. This corresponds to the peaks C and D, of the waves of induced E. M. F. shown in Fig. 64.

101. In practice, however, continuous-current generators do not produce this type of E. M. F. Fig. 68 represents, in cross-section, a common type of generator armature, situated between two field poles N and S. A type of generator, armature and field poles, similar to this, is seen in Fig. 1.

The flux from these poles passes readily into and out of the armature surface as indicated by the arrows. In other words,

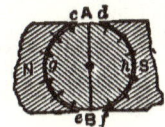

FIG. 68.—CROSS-SECTION OF BIPOLAR DRUM ARMATURE.

the flux cuts the surface of the armature at right angles, while, in the cases shown in Figs. 57 to 60, the conducting loop is only cut by the flux at right angles in two positions 180° apart, so that the curve of E. M. F. is peaked at these points, and descends rapidly from them on each side.

102. Suppose in Fig. 68 that the difference of magnetic potential, maintained between N and S, is 2,000 gilberts, that the diameter of the armature core g o h, is 40 cms., that its length is 100 cms., and that the air-gap or *entrefer* is 1 cm.; then, if the reluctance of the iron armature core be regarded as negligibly small, the magnetic potential between the polar surfaces and the armature surface on each side, that is between c N e and A g B, also between d S f and A h B, will be 1,000 gilberts. The magnetic intensity in the air may be obtained in two ways.

(1.) By considering the total reluctance of the air-gap and obtaining, by this means, the total flux. Thus the polar surface represented is 55 cms. in arc $\times$ 100 cms. in breadth $= 5,500$

sq. cms. The reluctance of the air-gap on either side of the armature is, therefore, $\frac{1}{5,500}$ oersted, and the total flux passing through the air will, therefore, be $\Phi = \frac{\mathfrak{F}}{\mathfrak{R}} = \frac{1,000}{\frac{1}{5,500}} = 5,500,000$ webers. This flux, divided by the area through which it passes, gives the intensity, or $\frac{5,500,000}{5,500} = 1,000$ gausses.

(2.) The magnetic intensity is, as we have seen (Par. 53), numerically equal to the drop of magnetic potential in air, or other non-magnetic material, per centimetre, so that the drop

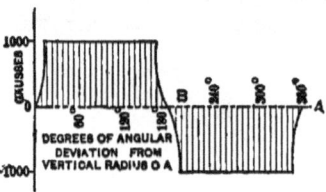

FIG. 69.—DIAGRAM OF MAGNETIC INTENSITY IN AIR-GAP.

of potential being here 1,000 gilberts in 1 cm. of distance in air, the intensity must be 1,000 gausses. Representing the intensity graphically, as shown in Fig. 69, it will be seen that the intensity is uniform from c to e, Fig. 68, and then descends rapidly to zero at B, where it changes sign and becomes negatively directed, and is then uniform from f to d, falling again to zero at A. The flux direction, therefore, changes sign twice in each revolution.

103. If a wire $A\,B$, be wound as a loop around the armature, it will, when the armature revolves, cut this flux at right angles, and will, therefore, have induced in it an E. M. F. which must be of the same type graphically as the curve in Fig. 69. Thus, if the surface of the armature moves at a rate of 50 cms. per second, the E. M. F. induced in the loop will be $2\,v\,l\,\mathfrak{B}$, the factor 2 being required, since both sides of the loop are cutting flux, one at A, and the other at B; or, $2 \times 50 \times 100 \times 1,000 = 10,000,000$ C. G. S. units = 0.1 volt.

except at the moment when the wires emerge from beneath the pole pieces. This curve is represented in Fig. 70, where the distance OF, represents the time of one complete revolution of the armature, and the elevation of A, corresponds to 0.1 volt. If the armature be set revolving at twice this speed, the time occupied in a revolution will be halved, but the E. M. F. being proportional to the rate of cutting flux, will

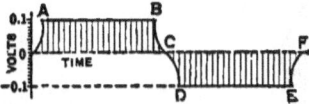

FIG. 70.—DIAGRAM OF INDUCED E. M. F. IN ARMATURE TURN.

be doubled, as represented in Fig. 71, where the E. M. F. is alternately 0.2 volt in each direction. By the aid of a suitably adjusted commutator, the E. M. F. instead of changing sign, can be kept unidirectional in an external circuit, following the curve $o\,a\,b\,c\,k\,l\,f\,g\,h\,j$.

104. We may regard the E. M. F. of the loop as being induced either by the cutting of the flux by the wire at the arma-

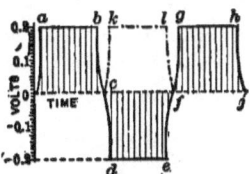

FIG. 71.—DIAGRAM OF INDUCED E. M. F. IN ARMATURE TURN AT DOUBLED SPEED OF ROTATION.

ture surface, or by the enclosure of the flux by the loop. The flux enclosed by the loop is represented by Fig. 72, where at the initial position at AB, the loop encloses 5,500,000 webers. As the armature is rotated counter-clockwise, so that A, is carried toward N, the flux enclosed by the loop diminishes, until, when it reaches the horizontal position, the flux through the loop is zero. As the rotation continues, the flux re-enters the loop in the opposite direction, and becomes 5.5 megawebers at a position 180° distant from the initial position AB. The rate of change of flux enclosed, or the gradient of the curve, shown in Fig. 72, is uniform, since the curve is uniformly steep, except near the position of maximum flux, where the gradient is considerably reduced, and the E. M. F. correspondingly reduced as already observed in Figs. 70 and 71.

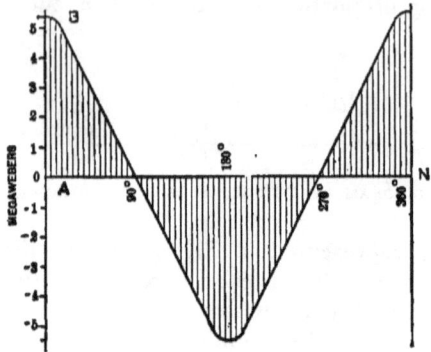

FIG. 72.—REPRESENTING DIAGRAM OF FLUX ENCLOSED BY LOOP OF ARMATURE.

105. When, however, the wire instead of being on the surface of the armature is buried in a groove in the iron, as in a toothed-core armature (Par. 22), and as shown in Fig. 73, it is often more convenient, for purposes of calculation, to consider the E. M. F. as due to enclosing, rather than to cutting flux. The following rule, will, therefore, be of assistance in

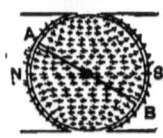

FIG. 73.—ARMATURE LOOP ROTATING IN BIPOLAR FIELD.

determining the direction of the E. M. F. induced in a loop. Bearing in mind the fact that a watch dial is visible, to an observer who holds it facing him, by the light which proceeds in straight lines from the watch to his eye, then the direction of the E. M. F. induced in the loop, regarded as the outline of the watch face, can be remembered by the following rule.

INDUCTION IN DYNAMO ARMATURES.   95

The E. M. F. induced in the loop has the same direction as the motion of the hands of the watch, when the flux entering the loop has the same direction as the light.

106. Flux entering the loop in the *opposite* direction, or from the observer, will induce an E. M. F. in the *opposite* direction to the hands of the watch, that is, counter-clockwise.

Emptying a loop of flux produces in it an E. M. F. in the opposite direction to that produced by filling it.

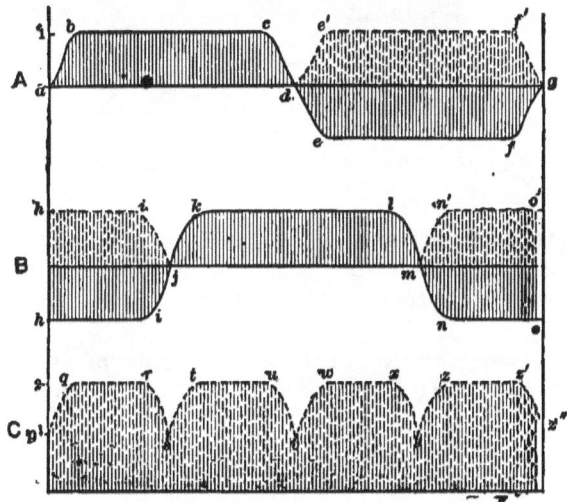

FIG. 74.—DIAGRAMS OF E. M. F.

107. Fig. 68, shows a single loop of wire wound upon a drum armature, which by its rotation in the flux, has an E. M. F. induced in it of the same type as is graphically represented in the curve of Fig. 69. Supposing that the speed of revolution is such as to produce an E. M. F. of say one volt, in this conducting loop, during its passage beneath the pole faces, then if two turns of wire be wound on the armature at right angles, as shown at AB and CD, Fig. 75, they will each generate E. M. F. of the same value, in their proper order, as they pass through the flux, and if the E. M. F. from AB, is represented by the curve of $a\ b\ c\ d\ e\ f\ g$, of Fig. 74 A, and the E. M. F. in the loop CD, be represented simultaneously by the curve of *h i j k l m n o*, of Fig. 74 *B*, then, by properly adding and co-directing the E. M. Fs. so produced, by the aid of a suitable commutator, we obtain an E. M. F. of two volts, as shown in Fig. 75, *C*, by the curve *p q r s t u v w x y z z' z''*. Moreover, while the E. M. F. produced from one wire alone

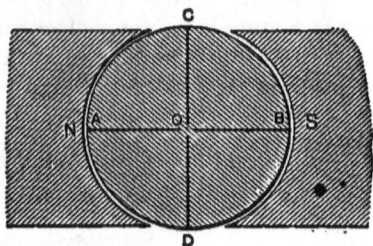

FIG. 75.—DRUM ARMATURE WOUND WITH TWO TURNS OF WIRE AT RIGHT ANGLES TO EACH OTHER.

fluctuates between 0 and 1 volt, four times per revolution, the E. M. F. produced by the combination fluctuates between 1 and 2 volts, eight times per revolution.

108. If now, instead of two loops being wound on the armature, there are six loops, as shown in Fig. 76, the E. M. F.

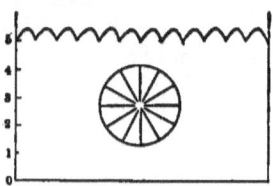

FIG. 76.—DRUM ARMATURE OF SIX EQUIDISTANT TURNS, WITH CORRESPONDING CURVE OF E. M. F.

generated in these, added and co-directed by the aid of a suitable commutator, will be represented by the curve in the same figure, and while the E. M. F. generated in any one of the conducting loops fluctuates between 0 and 1 volt, four times per revolution, the total E. M. F. produced under these conditions would vary between 5 and 5.5 volts, 24 times per revolution. In the same manner, if instead of 6 conducting loops being placed on the armature, there are 12 such loops, as shown in Fig. 77, the total E. M. F., if added and co-directed by a suitable commutator as before, would vary between 10.6 and 10.8 volts, 48 times per revolution, as shown by the curve.

109. An inspection of the preceding curves of E. M. F. will show that, while the total E. M. F. capable of being produced from a combination of conducting loops, is less than the sum of the maximum E. M. Fs. in each separately, yet their com-

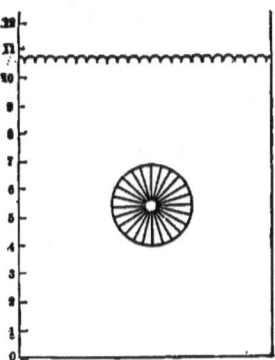

FIG. 77.—DRUM ARMATURE OF TWELVE EQUIVALENT TURNS, WITH CORRESPONDING CURVE OF E. M. F.

bined E. M. F. is much more nearly uniform than their separate E. M. Fs., and tends to become constant as the number of loops is increased, the curve of the total E. M. F. tending to become more and more nearly a horizontal straight line.

110. It must be carefully remembered that the E. M. F. generated in any single turn does not necessarily continue uniform during the passage of the turn beneath the pole; or, in other words, that the crests of the waves of E. M. F. are not necessarily straight lines, such as are indicated in Fig. 69, 70, 71, and 74. These crests will be straight lines, only if, as hitherto assumed, the intensity in the air-gap remains uniform over the entire polar surface. In practice this is rarely the case. The intensity may be either greater or less at the centre of the poleface than at the edges, but is usually greater, the flux tapering off toward the polar edges. This is owing to the fact that the reluctance in the magnetic circuit is usually a minimum, at or near the polar centre, with a consequent increase in intensity in that region. The same rules apply, however, even when the wave form of E. M. F., as generated by the wires singly, is complex. The effect of winding a number of turns around

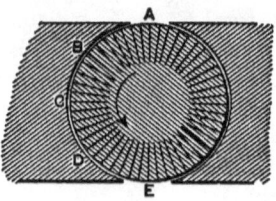

FIG. 78.—DRUM ARMATURE WITH TWENTY-FOUR TURNS IN BIPOLAR FIELD.

the armature, and uniting their E. M. Fs., is to produce an aggregate E. M. F. that is much more nearly uniform than the E. M. F. in each separate turn.

Thus Fig. 78 represents a drum armature with twenty-four complete loops, or forty-eight wires, lying over its surface and uniformly dispersed. If this armature be rotated in a bipolar field which is of such strength and distribution that each turn

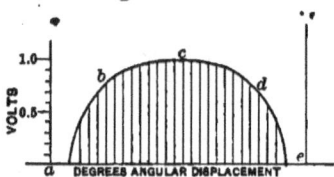

FIG. 79.—E. M. F. DIAGRAM OF ONE TURN ON ARMATURE.

has induced in it an E. M. F., such as is represented in Fig. 79, that is to say, no E. M. F. at the point a, about 0.7 volt at b, a maximum of about 0.95 volt at c, and no E. M. F. at e; then, if with the aid of a suitable commutator, these loops are connected together so as to unite their E. M. Fs. into two equal series, the E. M. F. of the machine as obtained from the brushes on the commutator is represented during half a complete rotation by the curve in Fig. 80, the corresponding points of which are marked $a\ b\ c\ d\ e$. It will be observed that there are twenty-four undulations in this curve, each undulation corresponding to the step between the entrance of each turn under the pole-pieces.

111. Moreover, the shape of the polar edges must necessarily influence the rise and fall of the E. M. F. induced in each separate wire as it passes beneath the pole. For example, if

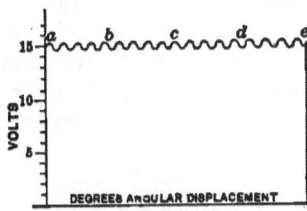

FIG. 80.—E. M. F. DIAGRAM OF ARMATURE COMBINING E. M. Fs. FROM SEPARATE TURNS.

the area of the pole-face be represented by the shaded area A in Fig. 81, the wires passing in succession beneath this pole, will have an E. M. F. induced first in a portion of their length, and finally throughout their entire length, so that the E. M. F. wave for each wire will rise gradually. If, however, the polar

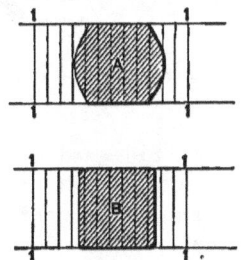

FIG. 81.—DIAGRAMS OF POLAR FACES OF DIFFERENT OUTLINE, OVER ARMATURE.

area be such as is represented at B, the wires enter the polar flux more suddenly, and the E. M. F. wave of each wire, at the beginning and end, will be rendered more abrupt. As regards continuous-current generators, there is but little advantage to be gained by variations in the shape of the pole-faces, since the aggregate E. M. F. of such a machine is rendered nearly uniform by the superposition of the E. M. Fs. in the various wires. Eddy currents in the conductors and iron core are, however, diminished by tapering the pole pieces, as at A.

112. In studying the arrangement of the wires on the surface of the armature in a generator, with the view of determining the E. M. F. generated by the revolution of the armature, it is necessary to observe that the E. M. F. developed does not depend directly upon the length of the armature wire which cuts magnetic flux, but does depend directly upon the amount of flux enclosed by the conducting loops during their revolu-

FIG. 82.—TYPE OF ARMATURE HAVING COMPARATIVELY LITTLE "IDLE" WIRE.

tion. It is a common error to regard all the wires on the free surface of an armature which do not pass through the magnetic flux as idle wires; and, consequently, detrimental to the efficient operation of the machine. This error comes from regarding the E. M. F. as produced alone by the cutting of flux, whereas in such a case, as for example, a pole armature (Fig. 17), none of the wire cuts the magnetic flux, and, consequently, would, by the preceding definition, be regarded as idle wire.

In reality, the generation of the E. M. F. is dependent on the embracing of flux by the loops, and since the so-called idle wire is necessary to form a part of the loop, it cannot properly be regarded as idle. It is, of course, to be remarked that in the event of the conducting loop having a fairly considerable part of its length formed of the so-called "idle" wire, in order to permit the loops to embrace a considerable amount of flux during their revolution, the rate of cutting flux by the parts that do cut, requires to be correspondingly increased, thus requiring a greater density of magnetic flux.

That this consideration is correct may be seen from an inspection of Figs. 82 and 83.

113. Fig. 82 represents a machine in which the armature is almost completely enclosed by polar surfaces, so that, even

FIG. 83.—TYPE OF ARMATURE HAVING COMPARATIVELY MUCH "IDLE" WIRE.

allowing for the free wire on the sides of the armature, sixty per cent. of the length of the wire is always in the magnetic flux, and forty per cent. is "idle." Fig. 83 shows a type of armature in which only about twenty-five per cent. of the length of the wire is at any time in the magnetic flux, so that about seventy-five per cent. is "idle." Yet, with equally advantageous circumstances as regards the cross-section of the iron core, speed of revolution, and the number of turns of wire, the E. M. F. from the machine shown in Fig. 83 is fully equal to, if not greater than, that developed in the armature of Fig. 82. If, for example, the polar surface in Fig. 82 were reduced by cutting it away along the lines *ab*, *cf* and *de*, thus removing the polar edges, and shortening the polar arc by about fifty per cent., the E. M. F. developed by the generator would not be reduced if the same total quantity of flux were forced through the armature as before. The change effected would be that the reluctance of the air-gap, between poles and armature on each side, would be increased, since the cross-sectional area of the air-gap would be diminished, and a greater M. M. F. would therefore be needed on the field magnets in order to produce the same flux through the circuit as before, but if this flux were reproduced, the amount enclosed with each turn of the armature by its revolution would be the same, and the total E. M. F. induced in the armature would be the same; or, regarding the question from a different standpoint, the intensity of flux in the air-gap would be increased about one hundred per cent., so that the wires would generate twice as much E. M. F. as before, but would only be generating E. M. F. about half the time in each revolution.

In other words, provided the armature core is traversed by a given magnetic intensity, it is a matter of indifference how much of its surface is covered by pole-pieces or how much left exposed with "idle wire," except as regards the amount of M. M. F. which will be needed to force the flux through the armature.

CHAPTER IX.

ELECTROMOTIVE FORCE INDUCED BY MAGNETO GENERATORS.

114. One of the earliest types of operative dynamos was that in which the field consisted of a permanent magnet, and the armature was of the Siemens, or shuttle-wound type. This armature consists essentially of a single coil of many turns of wire, wrapped in a deep longitudinal groove, formed on opposite sides of an iron cylinder. Owing to its simplicity, this early type of magneto-electric machine has survived in its competition with more advanced types, for such purposes as signal calls in telephony, and for firing electric fuses in mines. A machine of this type is shown in Fig. 84. The magnets M, M, are usually *compound ; i. e.*, consist of separate bars of hardened steel, with their like poles associated as shown in the side view. The magnets are thus combined to form a single magnetic circuit through the armature, by means of soft iron pole-pieces N' and S'. The armature core $A\ A$, was originally formed of a single piece of soft iron, but is now usually *laminated*, that is, formed of sheets of soft iron, laid side by side. The armature winding is in the form of a single coil or spool, and the ends of the coil are brought out to the insulated segments of the two part commutator $C\ C'$, Figs. 85 to 88.

115. In order to determine the E. M. F. capable of being produced by a generator of this type and of given dimensions, it is necessary first to ascertain the total quantity of flux which passes through the armature in the different positions it assumes during rotation. As shown in Fig. 85, the armature core lies at right angles to the polar line, and, consequently, no flux passes directly through its winding. When, during its motion, the armature reaches the position shown in Fig. 86, where the end A, has approached the north pole, the flux is threading through the armature in a direction from the north pole N, to the south pole S. In Fig. 87, the armature core is shown as lying directly between the pole-pieces. In this position the armature gives passage to the maximum amount of flux. In Fig. 88, the armature core is shown as moved beyond this position, and is now reducing the amount of flux threading through its core. Continuing rotation until the completion of a half turn, the position shown in Fig. 85, is reached, but now

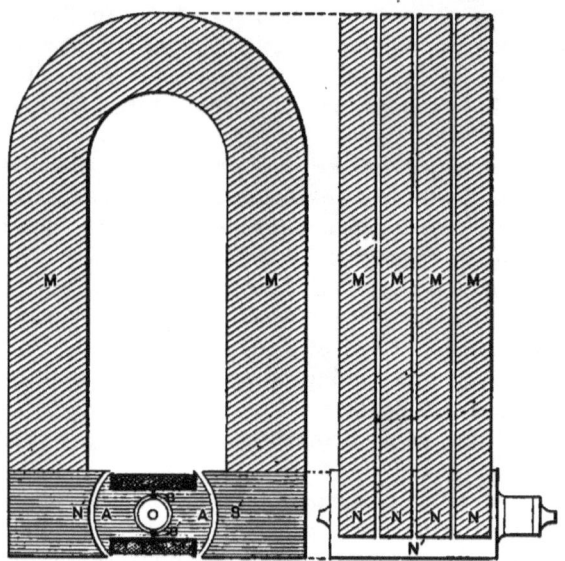

FIG. 84.—MAGNETO GENERATOR WITH SHUTTLE ARMATURE.

in the reverse direction; *i. e.*, with the end A, lowest instead of uppermost; and here the coil is emptied of flux as before.

116. It is evident, from a consideration of the preceding figures, that the amount of flux passing through the armature in any position depends upon the M. M. F. produced by the steel magnets; *i. e.*, upon their dimensions and shape, and on the reluctance of the air-gap, that is, on the dimensions and shape of the pole-pieces, as well as on the *entrefer* or air-gap lying between the poles and armature.

For practical purposes, a steel magnet may be regarded as producing a uniform difference of magnetic potential between its poles, except when the flux passing through the circuit represents an intensity greater than one kilogauss in the steel. We may practically consider that ordinary hard magnet steel maintains a permanent M. M. F. of 10 gilberts-per-centimetre of its length, independently of its cross-section, and at the same time possesses a reluctivity of $\frac{1}{150}$. If, then, the magnets shown in Fig. 84, are 30 cms. long and have a total cross-

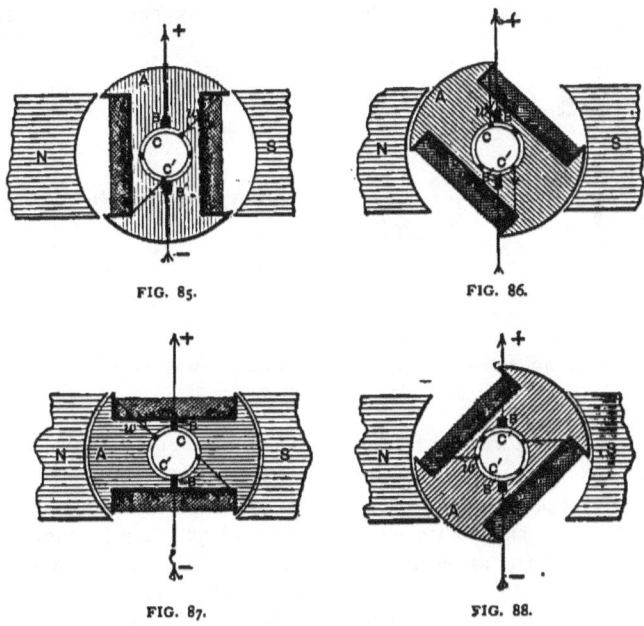

FIG. 85.   FIG. 86.

FIG. 87.   FIG. 88.

FIGS. 85, 86, 87, AND 88.—SHUTTLE-WOUND ARMATURE IN BIPOLAR FIELD.

section of 12 square centimetres, the M. M. F. they produce will be 300 gilberts, and their reluctance will be $\frac{30}{12} \times \frac{1}{150} = \frac{1}{60}$ oersted. Neglecting leakage, the flux which will pass through the armature will, therefore, be $\dfrac{300}{\frac{1}{60} + \mathcal{R}}$ webers, where $\mathcal{R}$, is the reluctance of the two air-gaps in series. If, then, we plot the total length of air space in cms. (twice the length of the air-gap), for different angular positions of the armature, and divide by the area of the armature beneath one pole in sq. cms., we obtain the reluctance ℛ, and, substituting its value in the above equation, we may determine, approximately, the magnetic flux through the armature for all positions during rotation.

117. Proceeding in this manner we obtain such a curve as is shown in Fig. 89, which represents the flux passing through

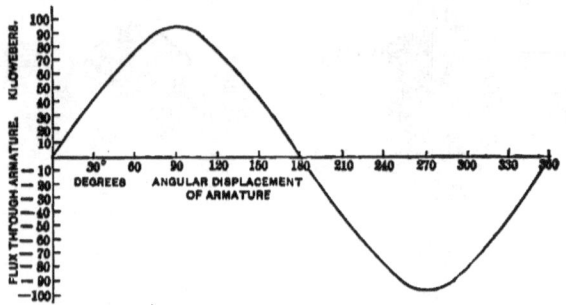

FIG. 89.—DIAGRAM OF FLUX PASSING THROUGH ARMATURE IN DIFFERENT ANGULAR POSITIONS.

the armature core at different positions of angular displacement from the initial position shown in Fig. 85, from actual measurements of a particular shuttle-wound machine of this type. An inspection of this figure will show that at 30° displacement the flux through the armature will amount to above 40 kilowebers, while at 90° displacement, the position of maximum flux, it will reach about 93 kilowebers. From this position the flux decreases until its value is zero at 180°, the position assumed by the armature when it has completed one half of a rotation and is again in the position represented in Fig. 85, but in the reverse direction. From this position onward, the direction of flux is reversed, the maximum flux being reached at an angular displacement of 270°, or ¾ of an entire rotation, completing a cycle at 360°.

118. Having thus obtained the value of the flux passing through the armature, it is a simple matter to determine the

E. M. F. at any speed of rotation; for, we have only to reconstruct the flux diagram of Fig. 89, to a horizontal scale of time in seconds, instead of angular displacement. This is shown in Fig. 90, for an assumed rate of rotation of 1.5 revolutions per second, or 90 revolutions per minute, the horizontal distance of *o m*, being taken as one second, and the vertical scale being taken for convenience smaller than in Fig. 89.

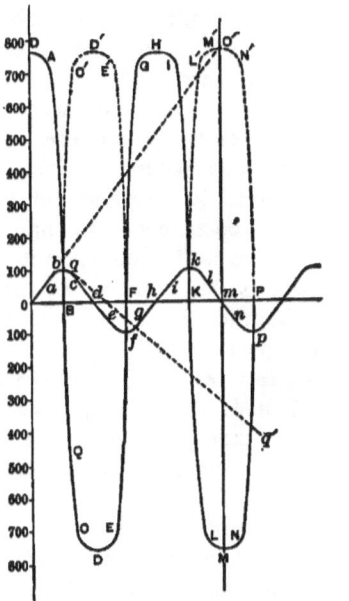

FIG. 90.—DIAGRAM OF FLUX PASSING THROUGH ARMATURE AT DIFFERENT PERIODS OF TIME.

The E. M. F. produced in any single loop or turn around the armature will be the rate of increase in the flux passing through the armature. If at the position *O*, commencing the curve, we continue the curve along the dotted tangent of *O O'*, for one second of time, we reach the ordinate *m O'*, of 770 kilowebers, and this is the rate at which flux is entering the loop at that moment; for, if the rate at *O*, were continued uniformly for an entire second, we should evidently reach the point *O'*. The E. M. F. existing at the moment of starting is, therefore, 770,000 C. G. S. units (of which 100,000,000 make one volt) or 0.0077 volt, and, if the number of turns around the armature core be 1,000, the total E. M. F. in the armature winding will be 7.7 volts. Again, if after a lapse of ⅙th of a second, the flux curve *o a b c d e f g h i k l m n*, be examined, it will be found that the curve has reached the point *b*, or its maximum positive value when it commences to descend toward *g*, so that the tangent is horizontal, representing that the rate of change of flux is zero, or similar to the condition of slack water in a tide-way. At this point, therefore, the E. M. F. in each turn on the armature is zero, and the curve of E. M. F. *O A B C D*, etc., touches the zero line at this point *B*.

Again at the point *q*, on the flux curve, if the change of flux were to continue for one second uniformly at this rate, we should follow the dotted line or tangent *q q'*, which reaches the ordinate —400, or 500 below *q'*, so that the rate of change at the point *q*, on the curve is 500 kilowebers, represented by the point *Q*, on the E. M. F. curve at that ordinate. Continuing in this way we trace the E. M. F. curve *O A B C D*, etc., showing that an alternating E. M. F. is produced in the armature, varying between +7.7 and —7.7 volts. At the rate of rotation assumed; namely, 1½ revolutions per second, there will be three alternations of E. M. F. per second, or twice the number of revolutions in that time.

119. Having now examined the means for determining the value of the E. M. F. developed in the armature, we will consider the effect of the commutator. It will be seen by reference to Figs. 85 to 88, the brushes *B*, *B'*, resting on the segments of the two-part commutator, that the direction of E. M. F. from the armature toward the external circuit is reversed at the moment when the core passes the position of maximum contained flux, as indicated by the change in the direction of the dotted loops *C' D' E'* and *L' M' N'*, relatively to the horizontal line. The E. M. F. generated by the armature as produced at the brushes *B*, *B'*, will be represented by the pulsating E. M. F., *O A B C' D' E' F G H I K L' M' N'*. It is evident that had we selected a higher rate of rotation, the E. M. F. of the machine would have been correspondingly increased.

120. The preceding considerations can only determine the value of the E. M. F. at the brushes, while the external circuit is open. As soon as the circuit of the armature is closed, the E. M. F. at the brushes is reduced, for the following reasons; viz.,

(1.) The current in the armature always produces an M. M. F., counter, or opposite to the M. M. F. of the field magnet, and, therefore, diminishes the flux through the magnetic circuit, thus causing a corresponding diminution in the value of the E. M. F. produced. Indeed, this opposing M. M. F. may, under certain circumstances, assume a magnitude sufficient to neutralize and destroy the permanent M. M. F. in the field magnets. This is one of the reasons why magneto generators are not employed on a large scale in practice.

(2.) The current through the armature produces in the resistance of the armature, a drop in the E. M. F. If, for example, the current through the armature at any instant be one ampere, and the resistance of the armature be 10 ohms, then in accordance with Ohm's law, the drop of E. M. F. produced in the armature, will be $1 \times 10 = 10$ volts.

(3.) The current through the armature not being steady, but pulsating, the variations in current strength will induce E. M. Fs. in the coil opposed to the change and, therefore, reducing the effective E. M. F.

CHAPTER X.

POLE ARMATURES.

121. The form of armature, which stands next in order of complexity to the shuttle-wound armature last described, is the *radial* or *pole armature*, represented in Figs. 91 and 92. Here the armature coils c, c, are wrapped, usually by hand, around radially extending laminated pole-pieces, formed from sheet iron punchings laid side by side. This type of machine is rarely found in continuous current generators, but is sometimes adopted in very small motors. The winding of such an armature is carried out as represented in Fig. 93, where the pole-pieces are shown at $P P$, and $P' P'$. Starting the winding at the point M, the coil A, is wound from A to B, as shown; the coil C, is then wound from B, through C to D; the coil E, from D, through E to F; the coil G, from F, through G to H; the coil J, from H, through J to K; the coil L, from K, through L to M, finally connecting the last end of the coil M, to the first end of the coil A, thus making the *closed-coil winding* shown in the figure. The connections of this winding to the six-part commutator will be seen from an inspection of the figure. The points M, B, D, F, H and K, are branches connected to the separate insulating segments of the commutator, brushes being provided in the position shown on a line connecting the centres of the pole-pieces. This commutator is shown in cross-section at P, Fig. 92. It will be seen that, owing to the conical boundaries of each armature coil, the winding is difficult to arrange. This type of generator is always operated by an electro-magnetic field.

122. Since the dimensions of machines with pole or radial armatures are always small, the reluctance of the circuit is practically wholly resident in the air spaces between the poles and armature projections, provided care be taken that the iron in the armature is not worked at an intensity above 10 kilogausses, or above 7 kilogausses in the field magnet, if the latter be of cast iron. If S, be the area of the polar face of a radial armature projection in square centimetres, and d, be the clearance or entrefer in cms., then $\dfrac{d}{S}$ will be the reluctance of the entrefer over each armature projection. Since there are four

FIG. 91.—POLE ARMATURE AT RIGHT ANGLES TO AXIS.

such air-gaps in multiple-series the total reluctance of the circuit provided in the case represented by Fig. 91, will be

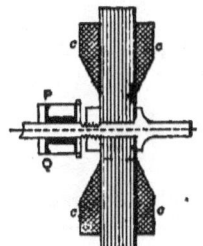

FIG. 92.—SECTION OF POLE ARMATURE THROUGH AXIS.

$\dfrac{d}{S}$ oersteds, assuming that the reluctance existing in the iron is neglected.

123. The distribution of the flux through the armature is diagrammatically represented in Fig. 95. If the cross-section of each armature core be s, square centimeters, then at no time will there be less than two radial projections carrying the total flux, and if 10 kilogausses be the limit permitted by the reluctance of the air-gap, the total flux to be forced through the armature will be $2 s \times 10{,}000 = 20{,}000\ s$, webers. The M. M. F. necessary on the field magnets will be $20{,}000 s \times \dfrac{d}{S}$ gilberts. For example, if $s = 1.3$ sq. cms., $d = 0.2$ cm., $s = 10$ sq. cms., the M. M. F. required will be $26{,}000 \times 0.02 = 520$ gilberts $= 416$ ampere-turns, and this must be the total excitation included on the limbs of the electro-magnet.

124. In order to determine the amount of flux passing through a single projection, let the armature be considered as slowly rotated counter-clockwise. Starting with the core 1,

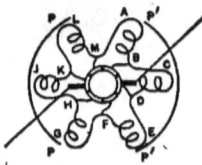

FIG. 93.—DIAGRAM SHOWING CONNECTIONS OF COIL WITH COMMUTATOR.

Fig. 95, the magnetic flux passing through it will be found by dividing half the M. M. F. by the reluctance of the air-gap over its face, or $\dfrac{260}{\frac{0.2}{10}} = 13{,}000$ webers. As it moves counter-clockwise towards 2, no appreciable change is effected in the amount of flux it carries, until the advancing edge of 2 emerges from beneath the polar face N_2. The flux through 1, rapidly diminishes until before 1 becomes halfway between the pole faces N_2 and S_1, it is entirely deprived of flux. When the position 3 is reached, the flux re-enters the coil of 1, but in the opposite direction, and when it passes position 3, the total maximum flux of 13 kilowebers is in the reverse direction. The curve, Fig. 94, commences at 13 kilowebers in the position corresponding to 1, Fig. 91, falls steadily from B to C, and, after a short pause, from C to D, where the coil lies midway between the poles, falls again from D to E, until the flux is 13 kilowebers negative, corresponding to the position 4. Continuing at this value to F, it rises to G, corresponding to the position 5, and then pauses at the zero line, in the gap between the poles, rising finally to J, corresponding to the original position 1, at K.

125. The E. M. F. established in any turn of the coil is found by ascertaining, from the speed of rotation, the rapidity with which the flux, threading through the coil, changes in value. If, for example, the armature be driven at a speed of 1,500

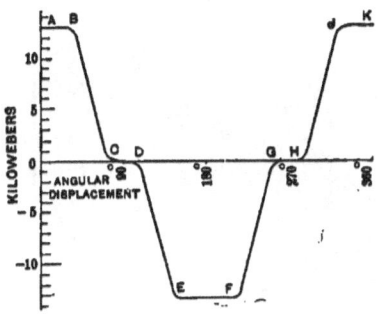

FIG. 94.—DIAGRAM SHOWING FLUX PASSING THROUGH ONE ARMATURE PROJECTION DURING A COMPLETE REVOLUTION.

revolutions per minute, or 25 revolutions per second, corresponding to the time of 0.04 second per revolution, the E. M. F. will evidently be zero at the positions represented by the straight line $A B$, $C D$, $E F$, $G H$, and $J K$ of Fig. 94, since here, the rate of change in the flux is practically zero, and the E. M. F. will be nearly uniform during the periods represented by $B C$, $D E$, $F G$, and $H J$, since the rate of change is nearly uniform in one direction or the other during those periods. As shown in Fig. 97, the E. M. F. in the single turn on the projection commencing at the position 1, is zero from o to b. From b, through b' to c, the flux diminishing at the rate of 13,000 webers in 0.00433 second, and, therefore, at the rate of 3,000,000 webers (3 megawebers) per second, and since 100 megawebers per second correspond to an E. M. F. of one volt, the E. M. F. in a single turn is —0.03 volt. Assuming 10 turns of wire on each armature projection, the total E. M. F. will be —0.3 volt at this period, and the ordinate bb, represents

—0.3 volt in Fig. 97. At $c'd$, corresponding to the position CD, Fig. 94, the E. M. F. is zero, falling again to —0.3 volt from d to e', corresponding to a change in flux from D to E, Fig. 94. After 0.02 second has elapsed, the E. M. F. reverses in direction and becomes positive, tracing the curve ff' gg' hh' jj'' k.

By the aid of the commutator, the E. M. Fs. in the coils, as soon as they change their direction, are reversed relatively

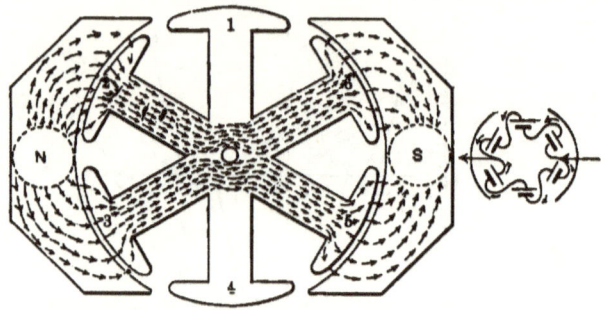

FIGS. 95 AND 96.—DISTRIBUTION OF FLUX AND E. M. F. AT POSITION SHOWN.

to the external circuit, and, therefore, preserve their direction externally, as can be seen by examination of Fig. 93.

126. We have thus far traced the E. M. F. as developed in a single polar projection, and so resulting from the variation of flux passing through it. During the time that the E. M. F. is being generated in this coil, a similar E. M. F. is being generated in the other coils, displaced, however, in time, by portions of a revolution. As shown in Fig. 96, the six coils on the armature have E. M. Fs. developed in them, being connected with the external circuit through the brushes in two parallel series, each of 3 series-connected coils. Each coil is, therefore, acting in its circuit for one half of a revolution before it is transferred to the opposite side, and while Fig. 97 represents the E. M. F. generated in any half revolution of one coil, we have to consider the E. M. Fs. coincidently being generated in its next neighbor on either side. This is shown in Fig. 98, where the E. M. F. of all three coils is developed independently on parallel lines one above the other, each E. M. F. being a repetition of that in Fig. 98, but displaced the ⅓th of a complete revolution. Fig. 99 represents

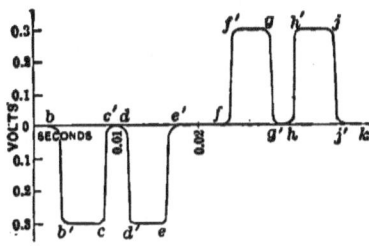

FIG. 97.

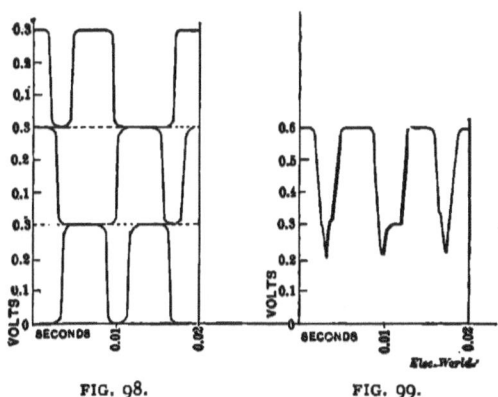

FIG. 98.   FIG. 99.

FIGS. 97, 98, AND 99.—E. M. F. WAVES GENERATED IN POLE ARMATURE.

the effects of combining or summing these three separately generated E. M. Fs. in the same circuit, and it will be seen that the E. M. F. pulsates between 0.2 and 0.6 volt.

127. If the resistance of the wire on each coil be r ohms, then the resistance of the three coils on each side of the armature will be $3\,r$, and the resistance of these two sides in parallel will, except at changes of segments, be $1.5\,r$, so that, neglecting the resistance of the brushes and brush contacts, the resistance of the armature will be $1.5\,r$ ohms.

The current strength which should be maintained by the generator, when on short circuit, would, therefore, reach $\dfrac{0.6}{1.5\ r}$ amperes, but in reality, the current will not reach this amount, owing, among other things, to the effect of self-induction in the armature, which, under load, tends to check the pulsations, and, consequently, renders them more nearly uniform, thus reducing the mean E. M. F.

CHAPTER XI.

GRAMME-RING ARMATURES.

128. The armature of the dynamo-electric machine which comes next in order of complexity, is that devised by Gramme, and now known generally as the Gramme-ring armature. This armature, as its name indicates, belongs to the type of ring armatures, and consists essentially of a ring-shaped laminated iron core wound with coils of insulated wire. In

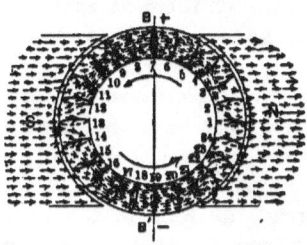

FIG. 100.—DIAGRAM OF GRAMME-RING ARMATURE IN BIPOLAR FIELD, TWENTY-FOUR SEPARATE TURNS.

the Gramme-ring armature shown in Fig. 100, the core is a simple ring of iron, wound with 24 separate turns of wire, placed so as to be able to revolve about its axis in the bipolar field N, S. Considering the ring to be first at rest, the turns 6, 7, 8, 18, 19 and 20 are represented as being linked with the total flux passing through the cross-section of the ring. If the total flux entering the armature at the north pole and leaving at the south pole, that is, passing from N to S, be two megawebers, then one megaweber passes through the upper half of the ring, and one megaweber through the lower half. The loops 5, 9, 17 and 21 are diagrammatically represented as having 900 kilowebers passing through them. The loops 4, 10, 16 and 22 carry 700 kilowebers; 3, 11, 15 and 23 carry 500 kilowebers; 2, 12, 14 and 24, 300 kilowebers; while 1 and 13 carry no flux.

129. Suppose now, the ring be given a uniform rotation of one revolution per second, in the direction of the large arrows. It is evident, that at any instant there is no change in the amount of flux linked with the turns occupying the positions 6, 7, 8, 18, 19 and 20; so that, although these contain a maximum amount of flux, they will have no E. M. F. generated in them. Loops 5 and 9, however, are in a position at which the flux they contain is changing; that is to say, the amount of flux that is passing through them at each instant has neither reached a maximum nor minimum; and the same is true with regard to the loops 17 and 21. In 5, the flux is increasing, and in 9, it is decreasing; consequently, the E. M. F. in 5 is directed oppositely to that in 9, and, according to rule, is indicated by the curved arrows (Par. 105); for, if coil 5 be regarded by an observer facing it from S, the flux, as the ring moves on, will thread the loop in the opposite direction to that of light coming from the face of the loop, considered as a watch dial, to the observer, and the E. M. F. generated in the loop will be directed counter-clockwise, while the E. M. F. in the loop 9 must have the opposite direction. Moreover, similar reasoning will show that all the coils to the left of the line $B B'$, that have E. M. Fs. generated in them, will have these E. M. Fs. similarly directed; *i. e.*, outwards, as shown, while all on the left-hand side of the line, will have the E. M. Fs. also similarly directed, but inwards. Loops 1 and 13, which lie parallel to the direction of the flux, will, in the position shown, have no flux threading through them, but during rotation, the rate of change of flux linked with them is a maximum; consequently, the E. M. F. induced in them is a maximum.

130. Instead of conceiving separate conducting loops to be wound on the surface of the armature, as shown in Fig. 100, let us suppose a continuous coil is wound on the surface of the armature as shown in Fig. 101, the first and last ends of the coils being connected together so as to make the winding continuous; then it is evident that the E. M. Fs. so acting being similarly directed on each side of the vertical line $B B'$, might be made to produce continuously an E. M. F. in the conducting wire. Moreover, if two wires, or collecting brushes, were employed in the positions B, B', the E. M. Fs. from the two halves of the ring would unite at the brushes B, B'.

Such a condition finds its analogue in the E. M. Fs. produced by two series-connected voltaic batteries connected as shown in Fig. 102, with their positive poles united at B, and their negative poles united at B'. The figure shows two batteries each of 9 cells connected in series. Here, as indicated, all the cells have equal E. M. F. This condition of affairs need not, however, exist in the Gramme-ring analogue, since the only requirement is that the sum of all the E. M. Fs.

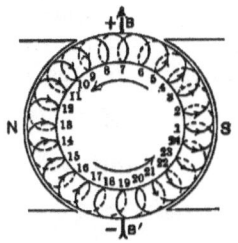

FIG. 101.—DIAGRAM OF GRAMME-RING ARMATURE IN BIPOLAR FIELD, TWENTY-FOUR SEPARATE TURNS.

generated in the coils on the right-hand side be equal to the sum of those on the left-hand side. In point of fact, as already observed, the E. M. Fs. are not the same in each of the coils, those at 1 and 13 having a maximum E. M. F., and those at 7 and 19 having zero E. M. F. Since these oppositely directed E. M. Fs. balance each other, no current will be produced in the armature unless an external circuit be provided, by joining the brushes B, B'.

131. Figure 100 shows no difference between the amount of flux threaded through the coils 6, 7 and 8; or 18, 19 and 20, and, consequently, according to theory, a total absence of induced E. M. F. in these coils. In practice, however, owing to leakage (Par. 77) and other causes, no coil is entirely free from having E. M. F. generated in it.

Moreover, the difference in the E. M. F. generated in coils 13, 12, 11 and 10, is not as great as might be inferred from their angular position on the armature, owing to the fact that

(Par. 100) the flux enters the armature core nearly uniformly all around its surface.

In order to determine the total E. M. F. generated in such an armature as is represented in Fig. 101, it is first necessary to determine the E. M. F. generated in a single turn. Let us consider a turn starting from the position 7, and therefore, generating no E. M. F., being carried by the uniform rotation of the armature in the direction of the arrows to the position 19, in a time t seconds. During this time the flux threading through it changes from $\frac{\Phi}{2}$ webers in one direction, to $\frac{\Phi}{2}$ webers in the opposite direction, and, therefore, the change in flux linkage will be Φ webers, Φ, being the total flux passing from N into S, through the armature. Whatever may be the distribution of flux through the armature, and in the air-gap, the average E. M. F. generated in the coil during this time will be $\frac{\Phi}{t}$ C. G. S. units of E. M. F. If the number of revolutions made by the armature per second be n, then one revolution takes place in the $\frac{1}{n}$th of a second, and a half revolution in the $\frac{1}{2n}$th of a second, so that $t = \frac{1}{2n}$, and the average E. M. F. is

$$\frac{\Phi}{\frac{1}{2n}} = 2n\,\Phi$$

132. If, for example, the armature be revolved at a speed of 600 revolutions per minute, or 10 revolutions per second, $n = 10$, and since Φ, has been assumed to be 2 megawebers, the average E. M. F. generated in any loop in passing from the position 7, to the position 19, will be $20 \times 2{,}000{,}000 = 40{,}000{,}000$ C. G. S. units, or 0.4 volt (Par. 82). The same E. M. F., oppositely directed, however, will exist on the average in any turn on the right-hand side of the line $B\,B'$. If the ring were wound with only four turns, say 1, 7, 13 and 19, the E. M. F. generated in these turns when placed in series and connected to the brushes B and B', would evidently fluctuate considerably; since, when the coils occupy the position shown, the E. M. Fs. would be a maximum in 1 and 13, and zero in 7 and 19, while, after ⅛th of a revolution, all four coils would be active. If, however, numerous turns are wound on the coil, it is evident that the total E. M. F. between the brushes B and B', will be very nearly uniform, since the only fluctuation which can take place is that coincident with the transfer of a single turn beneath the brush; consequently, in order to determine the total E. M. F. generated by the rotation of a Gramme-ring armature, it is only necessary to multiply the average E. M. F. in each turn by half the number of turns on the armature; *i. e.*, by the number

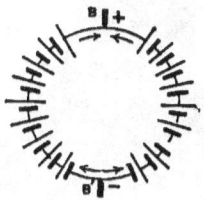

FIG. 102.—VOLTAIC ANALOGUE OF E. M. Fs. GENERATED IN GRAMME RING.

of turns active between B and B', on each side, so that if w, be the number of turns on the armature, counted once around, $\frac{w}{2}$ will be the number of turns active between brush and brush, and the total E. M. F. on each side of the armature will be

$$2\Phi n \times \frac{w}{2} = \Phi n w \text{ C. G. S. units} = \frac{\Phi n w}{100,000,000} \text{ volts.}$$

If $w = 24$, as in the case represented, then the total E. M. F. will be $2{,}000{,}000 \times 10 \times 24 = 480{,}000{,}000 = 4.8$ volts.

133. There is only one method, in practice, of connecting the separate coils of a Gramme-ring bipolar armature; namely, their continuous looping around the ring in a closed coil, as shown in Fig. 101.

Suppose that it is desired to utilize the generated E. M. Fs. for the purpose of supplying a current to an external circuit; it is then only necessary to apply suitable brushes, or conductors, at B and B', so as to rub continually against the external surface of the turns as they revolve, making the brushes sufficiently wide to maintain continuous contact.

Under these circumstances, during the rotation of the armature, a steady current will flow through the circuit maintained externally between B and B', B, being the positive pole of the machine, and B', the negative pole. Reversing the direction of the armature rotation will, of course, reverse the polarity of the brushes, as will also the reversal of the direc-

FIG. 103.—GRAMME-RING SEXTIPOLAR GENERATOR WITH BRUSHES COMMUTATING ON SURFACE OF ARMATURE.

tion of the magnetic flux. If, therefore, it be required to change the polarity of the brushes without changing the direction of rotation, it is only necessary to reverse the magnetic flux through the armature. Fig. 103 shows a Gramme-ring sextipolar generator, with the commutating brushes bearing directly on the metallic surface of the turns of conductor on the surface of the armature. This method, however, of commuting the current from a Gramme-ring armature is not the one in most frequent use; for, not only are the conductors upon the surface of the armature usually too small to bear brush friction without destructive wear, but also the relative amount of friction offered by brushes, placed upon so large a diameter, is considerable, except in the case

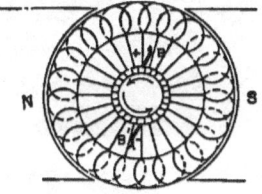

FIG. 104.—COMMUTATION OF CURRENTS FROM A GRAMME-RING ARMATURE BY A COMMUTATOR.

of very large machines. In order to avoid this, as well as for other reasons, it is usual to employ a special form of commutator, as represented diagrammatically in Fig. 104, where each turn is connected by a special conductor to a separately insulated segment of a commutator. This commutator, therefore, contains as many separate segments as there are turns on the

FIG. 105.—FORMS OF COMMUTATORS.

armature. Usually, however, there are many turns of wire on the armature to each segment of the commutator.

134. It is customary, in practice, to give a considerable length of free surface to the commutator bars, so as to increase the surface of contact and thus diminish the pressure that has to be applied. Fig. 105 shows two forms of such commutator. The separate segments are insulated from each other by mica strips. In order to provide for the connection of the wires from the armature to the separate commutator segments or bars B, metal projections or lugs L, attached to the bars, are provided. The bars, after being assembled, are held rigidly in place by the nut N.

Various forms of brushes are provided to maintain contact

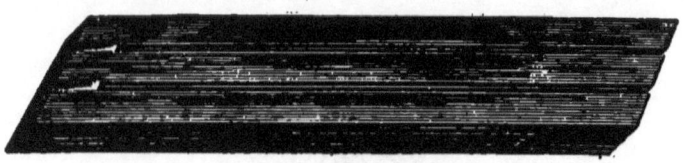

FIG. 106.—FORM OF GENERATOR BRUSH.

with the commutator bars. One form, consisting of wires and strips in alternate layers, is shown in Fig 106.

135. In the armature so far considered, it has been supposed that the condition as regards distribution of flux and the consequent generation of E. M. F. is symmetrical. It is possible, however, that in the construction of the machine this symme-

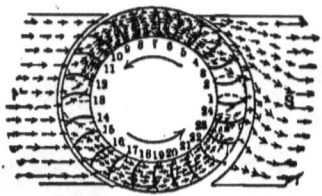

FIG. 107.—DIAGRAM REPRESENTING INFLUENCE OF MAGNETIC DISSYMMETRY.

try may not be secured. For example, in Fig. 107, the pole-piece S, is represented as being considerably further from the armature at its lower than at its upper edge, thereby increasing the reluctance of the air-gap at the lower edge, and producing *magnetic dissymmetry*, as represented by the distribution of flux arrows. It will be found, however, on examination, that despite this magnetic dissymmetry, the average E. M. F. produced in the coils would remain the same, although the distribution of this E. M. F. among the different turns necessarily varies. Thus if Φ, be, as before, the total flux through the armature, the lower half of the armature may take a certain fraction $n\Phi$, where n, is less than 0.5, while the upper half takes the balance $(1-n)\Phi$. The total change in flux linkage in passing from the position 7, to the position 19, will be $n\Phi - (1-n)\Phi = -\Phi$, as before, so that the average E. M. F. will not be altered by the dissymetry. It might be supposed, since the total flux passing through the armature remains the same, that no loss exists in an armature whose air-gap is thus widened, but a little consideration will show that the increased reluctance in the magnetic circuit necessitates a greater M. M. F. to drive the same amount of flux through

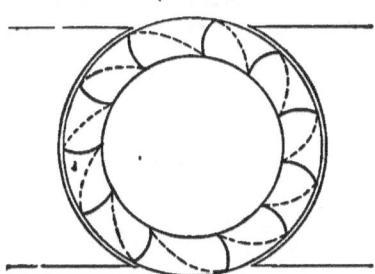

FIG. 108.—DIAGRAM REPRESENTING DISSYMMETRY OF WINDING.

the circuit, and, consequently, if the M. M. F. in the magnetic circuit remains the same, the total E. M. F. of the armature will be diminished. In addition to magnetic dissymmetry, a *dissymmetry of armature winding* may exist, such as shown in Fig. 108, where the right-hand half of the armature is seen to be wound with six turns while the opposite half is wound with five. In this case, supposing the armature to be rotating, there will be, at the moment represented, a greater E. M. F. in the right-hand half of the winding than in the left-hand half, and a current will therefore tend to flow through the armature under the influence of the resulting E. M. F., even when no external circuit is provided. When the armature has made half a revolution from the position shown, the left-hand half will be generating a greater E. M. F.; thus tending to force the current backward. Under these circumstances there will be produced in the armature an oscillating E. M. F., the number of oscillations in a given time being the same as the number of poles passed by any part of the armature in that time. That is to say, in a bipolar machine the frequency of the double oscillations will be equal to the number of revolutions of the armature per second. In a quadripolar machine it would be equal to twice the number of revolutions, and so on. These oscillations of current heat the armature winding and waste energy in it. Consequently, although symmetry is everywhere desirable in a machine, symmetry of armature winding is of greater importance than symmetry of magnetic flux distribution.

136. The armatures represented above are shown diagrammatically as rings of circular cross-section. In practice, how-

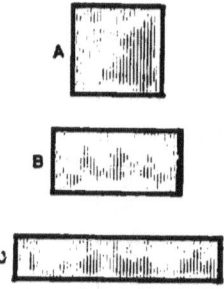

FIG. 109.—CROSS-SECTIONS OF GRAMME-RING ARMATURES.

ever, Gramme-ring armatures always have a rectangular cross-section, as represented in Fig. 109. We have seen that the E. M. F. of a Gramme armature, depends upon the number of turns of wire wound upon its surface, the flux passing through it, and the number of revolutions per second. The electric capability of a machine is expressed by $\frac{e^2}{r}$ (Par. 6); that is to say, its capability increases directly with the square of the E. M. F. and inversely with the resistance. For a given E. M. F. of the armature, it is, therefore, desirable to reduce the resistance as far as possible, in order to increase the electric capability of the machine. The shorter the length of the winding; *i. e.*, the shorter each turn, and the greater the cross-section of the wire, the less the resistance of armature. If R ohms be the resistance of all the wire on the armature, as measured in one length, then the resistance of a bipolar armature will be $\dfrac{R}{4}$ ohms, since two halves of the winding are in parallel; consequently, the resistance of the armature will depend upon the shape of its cross-section, since on this depends the length of each turn of conductor. A, B, and C, Fig. 109, represent the cross-sections of three different armature cores having the same area. Calling the length of one turn around A, unity, the length of a turn around B, will be 7 per cent. greater, and around C, 40 per cent. greater. Consequently, two armatures having respectively the cross-sections of A and C, and wound with the same size and number of turns of conductor, would have the same E. M. F., if driven at the same speed, when traversed by the same flux, but the armature C, would have 40 per cent. more resistance than the armature A, and its electrical capability would be about 30 per cent. less, $\left(\dfrac{1}{1.4}\right)$. It is, therefore, desirable in designing a Gramme-ring armature, to retain a nearly square cross-section. On the other hand, the section shown at C, offers for a given polar arc, a larger surface, and, consequently, a lower reluctance to the passage of the flux in the air-gap or entrefer, than in the case of the section A, so that it may be sometimes desirable to employ an armature of the type B, in order to reduce the air-gap reluctance, and, at the same time, not greatly to increase the length of winding.

It has been aptly remarked that a dynamo is a combination of compromises, since no single desideratum in its design can be completely realized.

CHAPTER XII.

CALCULATION OF THE WINDINGS OF A GRAMME-RING DYNAMO.

137. In order to show the application of the foregoing principles to the calculation of the E. M. F. produced in an armature of the Gramme type, we will take the case of a

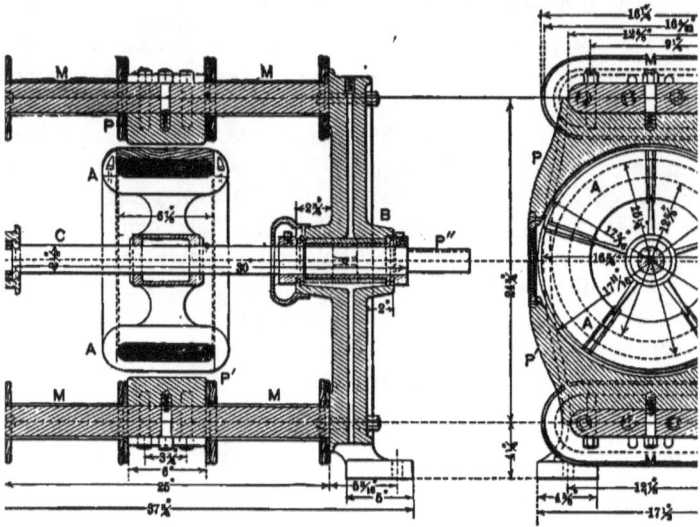

FIG. 110.—GRAMME TYPE ARC MACHINE.

bipolar Gramme-wound armature from dimensions given by Messrs. Owen and Skinner in a paper read before the American Institute of Electrical Engineers, May 16, 1894, to which paper the reader is referred for fuller particulars of construction and results.

Fig. 110, reproduced from the paper referred to, shows a vertical and a longitudinal cross-section of the machine, which is a bipolar, constant-current, Gramme-wound generator, of the Wood type, intended for the supply of any number of arc lamps in series up to 25, and, therefore, capable of supplying a total E. M. F. of approximately 1,200 volts at terminals, with a current strength of approximately, 10 amperes and an external activity of about 12 KW.

This machine, when complete, closely resembles the generator shown in Fig. 111. Referring to Fig. 110, the field magnet frame of cast iron is shown at M, M, M, M, the field coils being wound on spools and filling the spaces indicated. The shaft of the machine is supported in bearings B, B, and

FIG. 111.—GRAMME TYPE ARC MACHINE.

space is left on the shaft for a commutator, at C, and a driving pulley at P''. The bipolar field poles, produced by the M. M. F. of the magnet coils M, M, M, M, are shown at $P P, P' P'$. The Gramme-wound ring armature is shown at $A A A$. The dimensions of the machine are indicated in inches on the figures.

138. The field winding consists of 100 lbs. of No. 10 B. & S. gauge, single cotton-covered copper wire, the total resistance of the four coils in series being 15.75 ohms hot. The armature core is composed of soft charcoal iron wire of the cross-section shown. It is wound in 15 layers of No. 10 B. & S. gauge, and contains about 9,450 wires, each having a cross-section of 0.00817 square inch, or a total cross-section of 77.2 square inches = 498.1 sq. cms. The armature is wound in 100 sections of No. 14 B. & S. gauge, double cotton-covered copper wire, in 57 turns each, or 5,700 turns, making a total of 115 lbs. of wire, with a total resistance of 28.8 ohms hot, but which, being connected in two parallel halves, as represented in the figure, has a joint resistance between brushes of 7.2 ohms. Assuming 10 amperes to flow through the machine, the drop in the armature will be 72 volts, and the drop in the field magnets 157.5 volts, making the total drop in the machine 229.5 volts. When, therefore, the pressure at the machine terminals is 1,200 volts, the E. M. F. generated by the machine is practically 1,430 volts, or $1,430 \times 10^8 = 1.430 \times 10^{11}$ C. G. S. units of E. M. F.

139. The formula for determining the E. M. F. generated by a bipolar armature is

$$E = \Phi\, n\, w \qquad \text{C. G. S. units (Par. 132).}$$

Consequently, $\Phi = \dfrac{E}{n\, w}.$

The speed of this generator is stated to be 1,000 revolutions per minute, or 16.67 revolutions per second, and w, is 5,700, therefore, $\Phi = \dfrac{1.43 \times 10^{11}}{16.67 \times 5{,}700} = 1.505 \times 10^6.$ The total flux through the armature is, therefore, 1.5 megawebers.

140. Assuming that the M. M. F. required for this machine were not known, it could be calculated in the following way: We first determine the flux density in the various parts of the circuit, and from that the reluctivity and reluctance of the various portions.

The cross-section of the armature core, as already stated, is 498 sq. cms. and if the flux goes through each side or cross-section of the armature, the intensity in the armature is, therefore, $\dfrac{0.75 \times 10^6}{498} = 15{,}060$ gausses. The arc covered by each pole-piece is, approximately, 55 cms., and the effective breadth $6.5'' = 16.5$ cms., so that the area of the polar surface is, approximately, $55 \times 16.5 = 907.5$ sq. cms. The total flux passes through this surface, and the mean intensity in the air-gap is $\dfrac{1.5 \times 10^8}{907.5} = 16.58$ gausses.

141. Fig. 112 represents diagrammatically the arrangement of magnetic circuits through the machine, where M, M, M, M, represent the field magnet cores, P, P' the pole-pieces and $A\,A$

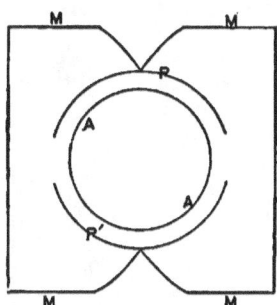

FIG. 112.—DIAGRAM OF MAGNETIC CIRCUIT.

the armature. Fig. 113, represents diagrammatically the voltaic analogue of the magnetic circuits, where $M_1 M_2 M_3 M_4$ are four batteries, whose E. M. Fs. correspond to the M. M. Fs. of the field-magnet coils. M_1 and M_2, form one circuit through the field frame, a certain mean length of the pole-pieces, and a mean length in the armature a, together with the two resistances $R_1 R_2$ in the air-gaps. A similar circuit is provided for the E. M. Fs. M_3 and M_4, through the air-gap resistances R_3 R_4, and the mean lengths of armature and pole-pieces. The equivalent arrangement of circuits is represented in Fig. 114, where M, M, are E. M. Fs., each equal to M, in the preceding figure, while the resistance of the double circuit through the field frame is one half of that of either of the resistances represented in Fig. 113.

142. The flux through the field cores will be greater than the flux through the armature by reason of a certain leakage which occurs over the surface of the magnetic circuit. This leakage is represented diagrammatically in Fig. 113, as taking place in a branch circuit or dotted semi-circle around the field coils, but, in reality, the leakage takes place in an extended system of branched or derived circuits between the polar surfaces and portions of the entire field frame. The calculation of the various reluctances in the air-path offered to leakage is very complex, and it is preferable, rather than to attempt such calculation, to refer to experimental data already acquired with machines of similar type. The *leakage factor*, or the ratio of total flux through the field magnet cores to the total flux pass-

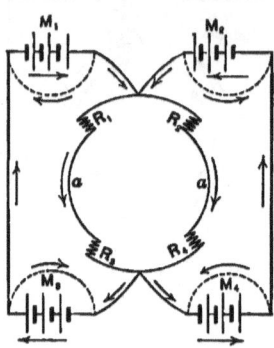

FIG. 113.—VOLTAIC ANALOGUE OF MAGNETIC CIRCUIT.

ing from them through the armature, for a machine of this type, is approximately 1.7; so that, since the useful flux passing through the armature from each circuit $M_1 M_3$ and $M_2 M_4$, Fig. 113, is 0.75 megaweber, the flux through the field cores may be taken as $0.75 \times 1.7 = 1.275$ megawebers. The cross-section of the cores is found to be 176.8 sq. cms., so that the intensity in them is, approximately, $\dfrac{1.275 \times 10^6}{176.8} = 7{,}211$ gausses.

143. The reluctivity of the soft wrought iron armature at a density of 1.5 kilogausses, is, approximately, 0.0045 (Fig. 47), the mean length of the flux paths through the armature 38 cms., and the cross section 498 square cms. The reluctance of each side of the armature a, Figs. 113 and 114 is, therefore, $\dfrac{38 \times 0.0045}{498} = 0.000343$ oersted. The joint reluctance of the armature will, therefore, be 0.00017 oersted; and, since the armature does not consist of continuous sheets of iron, but of wires, and the flux has to penetrate from wire to wire downward through small air-gaps, the total effective reluctance of the armature will be approximately 0.001 oersted. The length of the air-gap or entrefer, is $1.22'' = 3.1$ cms., and the area as already determined, 907.5 sq. cms. so that the reluctance in each air-gap will be $\frac{3.1}{907.5} = 0.003416$ oersted, the total reluctance in the air, as seen in Fig. 110, will then be 0.006832 oersted.

The reluctivity of the cast iron in the field frame at a mean intensity of 7,211 gausses, may be taken as 0.009 (Fig. 47). The length of the mean path in the field on each side of the machine is, approximately, 152.4 cms., and its cross-sectional area 176.8 sq. cms.; so that the reluctance in each half of the field will be, approximately, $\frac{152.4}{176.8} \times 0.009 = 0.00776$ oersted. The total flux being divided between the two sides of the field, the joint reluctance, as represented in Fig. 114, will be 0.00388 oersted.

		Gilberts.
The drop of magnetic potential in the reluctance of the armature (ΦR) will be,	$1.5 \times 10^6 \times 0.001 =$	1,500
The drop of magnetic potential in the reluctance of the the air,	$1.5 \times 10^6 \times 0.006832 =$	10,248
The drop of magnetic potential in the reluctance of the field,	$2.55 \times 10^6 \times 0.00388 =$	9,894
Total		21,642

Since one gilbert = 0.7854 ampere-turn, the total M. M. F. in the circuit will have to be very nearly 17,000 ampere-turns, or 8,500 ampere-turns on each of the spools M, M, M, M.

144. The preceding calculation is open to errors from several sources in the absence of definite experimental data, namely:

(1.) The assumed leakage factor may be inaccurate.

(2.) The mean lengths of the flux paths in various portions of the circuit may be inaccurate.

(3.) The assumed increase in the reluctance of the armature

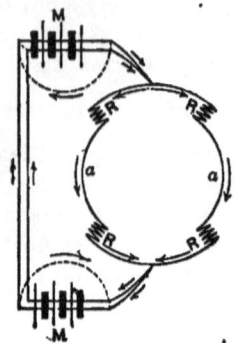

FIG. 114.—DIAGRAM OF VOLTAIC ANALOGUE.

due to its being formed of wires instead of solid sheets may be inaccurate.

(4.) The reluctivity of the cast iron employed in the machine may not be that of the sample of cast iron assumed.

In this, as in all constant-current machines, means are provided for maintaining a nearly constant current strength in the circuit, despite changes in the load, but a consideration of such means, and of the requirements of the magnetic circuit to permit such regulation, will preferably be postponed until armature reaction has been studied.

CHAPTER XIII.

MULTIPOLAR GRAMME-RING DYNAMOS.

145. A given type of bipolar Gramme machine having proved satisfactory as regards efficiency, ease of running and cost, at a full-load output of say 10 KW, it may have to be determined whether it would prove advantageous to maintain the same design for a machine of a greater output, say 80 KW. Let us assume that the linear dimensions of the 10-KW machine are doubled, with the same speed of revolution, say 1,000 revolutions per minute, maintained in the larger machine. Then, assuming the same magnetic intensity in the armature, the electromotive force will be four times as great, since the area of cross-section of the armature, and, consequently, the total useful flux, will be increased fourfold. The resistance of the armature will be halved; for each turn, though twice as long, will have a cross-sectional area four times greater.

The electric capability of the smaller machine being expressed by $\frac{e^2}{r}$ (Par. 6), that of the greater will be $\frac{(4e)^2}{\frac{1}{2}r} = 32\frac{e^2}{r}$, or 32 times greater than in the 10-KW machine; and, if the same relative efficiency is maintained in the larger machine the output will be 32 times greater. The weight of the larger machine would, of course, be eight times that of the smaller, and the output per pound of weight would, therefore, be four times greater in the larger machine. In reality, however, such a result is impracticable, as will now be shown.

146. Dynamo machines are either *belt-driven* or *direct-driven*. In the case of direct-driven generators, the speed of the generator is necessarily limited by the speed of the engine, and this, for well-known constructive reasons, has to be maintained comparatively low, and the larger the generator the slower the speed of rotation that has to be practically adopted.

Thus, while a 100-KW generator is commonly driven direct from an engine at a speed of about 250 revolutions per minute, a 200-KW generator is usually direct driven at about 150, and a 400-KW generator at about 100 revolutions per minute. In the case of belt-driven generators, the speed of belting is usually limited, except when driving alternators, to about 4,500 feet per minute; and, since larger generators require larger pulleys, their speed of rotation has to be diminished. While no exact rule can be applied for determining their speed, yet roughly, in American practice, the speed varies inversely as the cube root of the output, so that, when one generator has eight times the output of another of the same type, the speed of the greater machine would roughly be half that of the smaller.

If no other limitation existed besides efficiency, the effect of doubling the linear dimensions of any generator, even taking the reduced rotary speed into account, would result in producing about sixteen times the output for eight times the total weight; but large machines must necessarily possess a higher efficiency than small machines, not only owing to the fact that they would otherwise become too hot, the surface available for the dissipation of heat only increasing as the square of the linear dimensions, while the weight and quantity of heat increase as the cube of the dimensions,—but also because large machines are expected to have a higher efficiency from a commercial point of view.

147. Taking into account, therefore, the reduced rotary speed of larger machines, their limits of temperature elevation, and their necessity for an increased efficiency, the output only increases, approximately, as the cube of their linear dimensions; and, consequently, the output of the larger machine, per pound of weight, remains practically the same as that of the smaller. The output of belted continuous-current generators is commonly six watts per pound of net weight, and of direct-driven multipolar generators about eight watts per pound of net weight.

148. We have already seen (Par. 132) that the E. M. F. generated by a Gramme-ring armature, is $\Phi n w$ C. G. S. units, or $\Phi\frac{nw}{10^8}$ volts, and the resistance of the armature will be $\frac{R}{4}$ ohms, if R, be the resistance of the winding measured all the way round. Suppose now, that instead of employing a bipolar machine, we double the number of poles and produce a four-pole or quadripolar machine, as shown diagrammatically in Fig. 115. If we employ the same total useful flux Φ, through each pole, the average rate of change of flux through the turns on the armature will be doubled, since the flux through any turn is now completely reversed in one-half of a revolution,

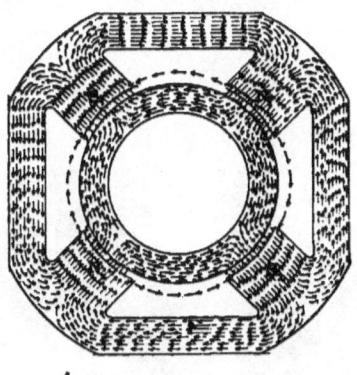

FIG. 115.—DIAGRAM OF MAGNETIC CIRCUITS IN QUADRIPOLAR GRAMME GENERATOR.

instead of in one complete revolution as before. The average E. M. F. in each turn will therefore be doubled. In Fig. 115 the magnetic circuits of a quadripolar Gramme generator are shown diagrammatically by the flux arrows. Here, as will be seen, four distinct magnetic circuits exist through the armature, instead of the two which always exist in the armature of a bipolar generator. In this type of field frame four magnetizing coils must be used. These may be obtained in one of two ways; namely,

(1.) By placing the magnet coils directly on the field magnet cores, as shown in Fig. 116; or,

(2.) By placing one coil on each yoke, as represented in Fig. 117.

149. In the same way, if we employ a field frame with six magnetic poles, as shown in Fig. 118, the flux will be reversed through each turn of wire three times in each revolution, and, consequently, the average E. M. F. in each turn will be increased threefold over that of a bipolar armature. In Fig. 118 there are six magnetic circuits through the armature. Considering any segment of the armature underneath a pole

FIG. 116.—QUADRIPOLAR GENERATOR WITH GRAMME ARMATURE.

as, for example, between n_2 and p, the turn occupying the position at n_2, is filled with flux in an upward direction. As the armature advances in the direction of the large arrows, the flux through this turn will be diminished, and, when it reaches the middle of the pole piece S_2, it will be completely emptied of flux. The E. M. F. in the loop, during this portion of the revolution, will be directed outward on the ring, as shown by the double-headed arrows. After passing the centre of the pole piece S_2, the flux through the loop begins to increase, but

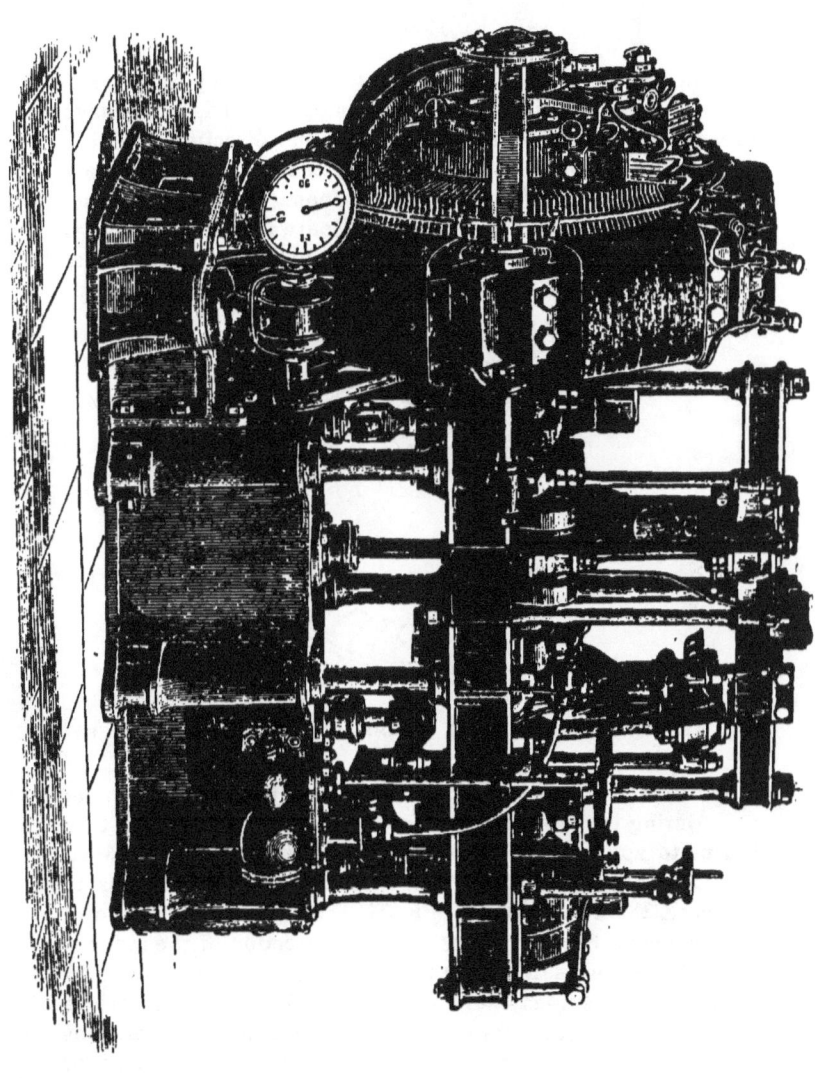

now in the opposite direction, the flux passing downward through the loop instead of upward as before, and, as we have already seen, flux entering a loop in one direction produces the same direction of E. M. F. around the loop as flux oppositely directed withdrawing from the loop (Par. 105). Consequently, the E. M. F. is still directed outwards on the ring, as indicated by the double-headed arrows, until the turn reaches the position p_1. In other words, the E. M. F. in a loop is simi-

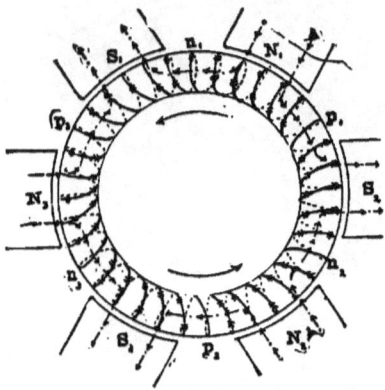

FIG. 118.—DIAGRAM OF SIX-POLE GRAMME-RING ARMATURE AND E. M. FS.

larly directed during its motion toward and from the same pole; *i. e.*, during its passage past a pole. When, however, the turn begins to approach the pole N_1, after being completely filled with the downward flux at p_1; *i. e.*, as the flux in it begins to decrease, the direction of the E. M. F. in it reverses, as shown by the double-headed arrows, and this direction of the induced E. M. F. continues until the turn reaches the position n_1. By tracing the directions of the induced E. M. Fs. in the various turns of the ring, as shown, it will be seen that the positions p_1, p_2, and p_3, are points at which the E. M. F. is positive, or directed outwards, while the positions n_1, n_2, and n_3, are points at which the E. M. F. is negative, or directed inwards. There will be no current passing through the armature in the condition represented, if the winding of the armature be symmetrical, since the E. M. Fs. in the various segments must be equal and opposite. If, however, brushes be applied to the surface of the armature at the positions p_1, p_2, p_3, and n_1, n_2, n_3, any pair of these, including one positive and one negative brush, will be capable of supplying a current through an external circuit.

150. When, therefore, an ordinary Gramme-ring winding is employed, there will be one brush placed between each pair of poles, or, in all, as many brushes as there are poles. Fig. 119

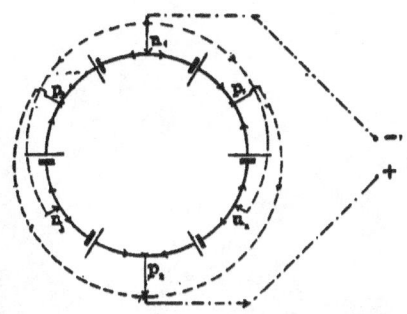

FIG. 119.—DIAGRAM OF CONNECTIONS BETWEEN BRUSHES OF A SIMPLE GRAMME-RING WINDING OF A SEXTIPOLAR ARMATURE.

represents the connections employed to unite the various segmental E. M. Fs. The E. M. F. of the armature is equal to that of one of its segments, but the resistance of the armature is inversely as the number of segments and poles, and if R, be the resistance of the entire armature winding, $\frac{R}{p^2}$ will be the joint resistance between brushes, for there will be p sections in parallel, each of which will have $\frac{R}{p}$ ohms. Consequently, in a six-pole armature, there will be six segments in parallel, each having a resistance of $\frac{R}{6}$, making the joint resistance $\frac{R}{36}$, or $\frac{R}{6^2}$.

Fig. 120 represents the mechanical arrangement for rigidly supporting the armature of a direct-driven octopolar Gramme-ring generator with eight sets of brushes pressing upon one side of the armature, thus dispensing with the use of a separate commutator. The central driving pulley PPP, supports upon its arched face two rings R, R'. These rings clamp between them the armature core, and are clamped together by 14 stout bolts. Where the supports ss, interfere with the winding of the conductor inside the armature, the conductors are carried on the supports as at $a\ b\ c$ and d.

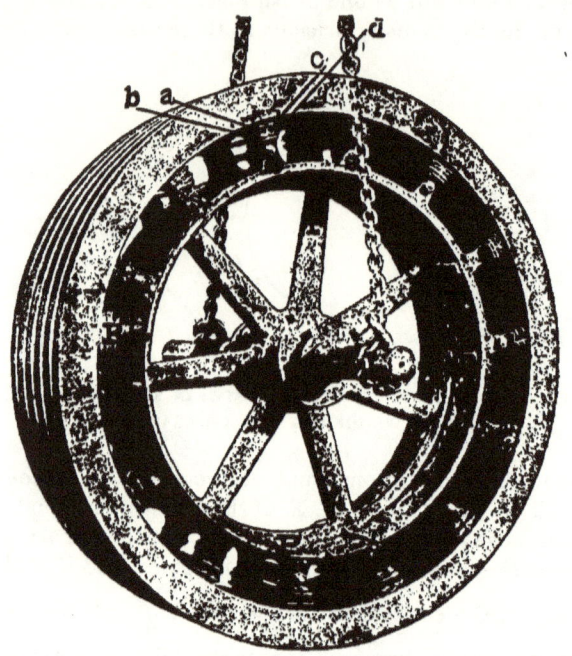

FIG. 120.—GRAMME-RING MULTIPOLAR ARMATURE.

151. It is not absolutely necessary, however, to employ six brushes in a sextipolar machine; for, since in a machine of this type the three separate circuits are connected in parallel, connections may be carried within the armature between the various segments, permitting of the use of a single pair of brushes. Thus Fig. 121 represents a Gramme-ring armature, wound for a sextipolar field, with triangular cross-connections between its turns. In this case, the corresponding points p_1, p_2, p_3, and n_1, n_2, n_3, of Fig. 118, instead of being connected together by brushes externally as in Figs. 119 or 120, are connected together by wires internally. It is not, of course, necessary that every turn on the armature should be so cross-connected, but that the coils or group of turns which are led to the commutator should be cross-connected, so that each of the 36 turns, shown in Fig. 121, may represent a coil of many turns. Although the brushes are shown in Fig. 121, as being placed on

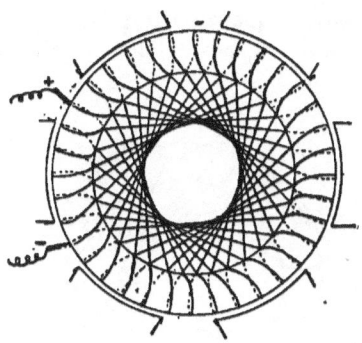

FIG. 121.—ARMATURE CROSS-CONNECTIONS FOR A SEXTIPOLAR GRAMME-RING WITH TWO BRUSHES.

adjacent segments, yet they may be equally well placed diametrically opposite to each other.

Fig. 122 represents the corresponding cross-connections for a quadripolar Gramme generator, employing a single pair of brushes. The advantage of cross-connections is the reduction in the number of brushes. The disadvantage of cross-connections lies in the extra complication of the armature connections. In large machines it is often an advantage to employ a number of brushes in order to carry off the current effectively.

152. Fig. 123 is a representation of a sextipolar generator whose magnetic field is produced by three magneto-motive forces, developed by coils placed as shown. The flux paths are represented diagrammatically by the dotted arrows at A. Each M. M. F. not only supplies magnetic flux through the segment of the armature immediately beneath it, but also contributes flux to the adjacent segments in combination with the neighboring M. M. Fs.

153. From the preceding considerations it is evident that while it is possible to design a bipolar generator for any desired output, yet, in practice, simple bipolar generators are not employed for outputs exceeding 150 KW, and, in fact, are seldom employed for more than 100 KW, since their dimensions become unwieldy and their output, per pound of weight, smaller than is capable of being obtained from a well-designed multipolar machine.

In the same way, a quadripolar generator can be made to possess any desired capacity; but, in the United States,

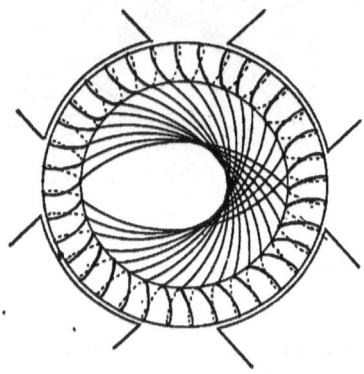

FIG. 122.—CROSS-CONNECTIONS FOR QUADRIPOLAR GRAMME-RING WITH TWO BRUSHES.

practice usually increases the number of the poles with an increase in the output of the machine. Thus, it is common to employ a four-pole or six-pole generator for outputs of from 25 to 100 KW, and 8 to 12 poles for a generator of 400 KW, capacity.

154. Should the armature of a multipolar generator not be concentric with the *polar bore ; i. e.*, if it is nearer one particular pole than any of the others, the reduction in the length of the air-gap opposite such pole, will reduce the reluctance of that particular magnetic circuit, and by reason of the increased flux through the armature at this point, induce a higher E. M. F. in the segments of the armature adjacent to the pole than in the remaining segments. If the armature be not *interconnected*; *i. e.*, if it employs as many pairs of brushes as there are poles, these unduly powerful E. M. Fs. can send no current through the armature as long as the brushes remain out of contact with the conductors; for an inspection of Figs. 118 and 119 will show that no abnormal increase of E. M. F. can exist in a single segment, but must be simultaneously generated in adjacent segments, and that such pairs of E. M. Fs. will counterbalance each other. When, however, the brushes are brought into contact with the armature conductors, thereby bringing the various segments into multiple connection with

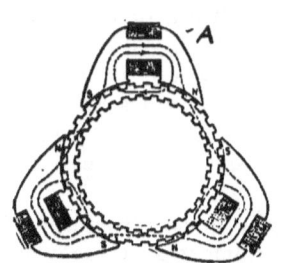 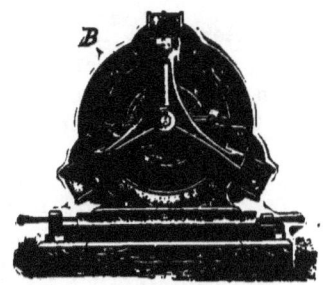

FIG. 123.—SEXTIPOLAR GRAMME-RING SUPPLIED BY THREE MAGNET COILS.

one another, a tendency will exist for the more powerful E. M. F. to reverse the direction of current through the weaker segments.

155. Whether this tendency will result in an actual reversal of current depends upon the difference of E. M. F. between the segments, their resistance, and the external resistance or load.

Let A and B, Fig. 124, represent the E. M. Fs. of any two segments in a multiple-connected Gramme-ring armature, and let the E. M. F., E, of A, be greater than the E. M. F., E', of B. Owing to drop of pressure in the internal resistance r, the pressure e, at the terminals p, q, will be less than the E. M. F., E, of the stronger segment A. If e, is greater than E', a current of $\dfrac{e-E'}{r}$ amperes will pass through the segment B, in the direction opposite to that in which its E. M. F. acts. If e, be equal to E', there will be no current through the segment B, while if e, be less than E', a current will be sent through B, in the direction in which its E. M. F. acts, but of strength less than that supplied by segment A. Thus, in Fig. 125, the E. M. F., E, of the stronger segment A,

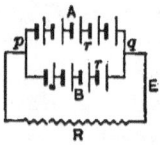

FIG. 124.—DIAGRAM OF E. M. FS. IN ADJACENT ARMATURE SEGMENTS.

is represented by the ordinate $e + d$. Owing to the resistance r, in the segment A, a drop of pressure d, will take place within it, and the pressure at its terminals will be e volts. If E' be less than e, the stronger segment A, will send a current back through the segment B, while if E', be greater than e, both segments will contribute current through the external load resistance R ohms.

For example, a separately-excited quadripolar generator of say 100 KW capacity, supplying 1,000 amperes at 100 volts

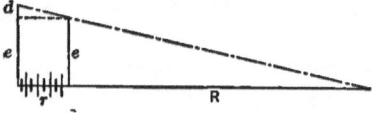

FIG. 125.—DIAGRAM OF E. M. FS. IN ADJACENT ARMATURE SEGMENTS.

terminal pressure, has a resistance in each of its four armature segments A, B, C, D, of $\frac{1}{100}$th ohm; then, provided its four magnetic circuits are balanced or equal, the full load on each segment will be 250 amperes, and the drop in each 2.5 volts; so that the four E. M. Fs. will be, Fig. 126 :

		E. M. F. Volts.	Drop. Volts.	Current. Amperes.	Wasted Power. Watts.
A	=	102.5	2.5	250	625
B	=	102.5	2.5	250	625
C	=	102.5	2.5	250	625
D	=	102.5	2.5	250	625
				1,000	2,500

The power expended in each segment of the armature by the current as I^2R, will be $\frac{250 \times 250}{100} = 625$ watts, and the total I^2R loss in the armature, 2,500 watts.

156. Considering one of the segments, say C, as normal, and that A, owing to the magnetic dissymmetry, gives an E. M. F. two volts in excess; B, one volt in excess; and D, one volt in

FIG. 126.—DIAGRAMMATIC ARRANGEMENT OF E. M. Fs. IN THE SEGMENTS OF A QUADRIPOLAR ARMATURE.

deficit; the excitation necessary for 1,000 amperes total output will produce (Fig. 127) the following conditions; namely,

		E. M. F. Volts.	Drop Volts.	Load. Amperes.	Wasted Power. Watts.
A	=	104	4	400	1,600
B	=	103	3	300	900
C	=	102	2	200	400
D	=	101	1	100	100
				1,000	3,000

157. The effect of magnetic dissymmetry in the segments, under the assumed difference of three volts, will produce, at

FIG. 127.—DIAGRAMMATIC ARRANGEMENT OF E. M. Fs. IN THE SEGMENTS OF A QUADRIPOLAR ARMATURE.

full load, a difference of output among the segments, ranging from 100 to 400 amperes, while the total power wasted in the armature winding will be increased 20 per cent.; namely, from 2,500 to 3,000 watts. The armature will, therefore, be raised to a higher temperature, owing to the magnetic dissymmetry, but this increase in temperature will not be localized, since, although at one moment a greater amount of heat is being produced in certain segments than in others, yet, owing to the rotation of the armature, the portions of the armature constituting these segments are constantly changing.

158. Suppose now the external circuit be entirely removed, the brushes remaining in contact with the conductors (Fig. 128)

FIG. 128.—DIAGRAMMATIC ARRANGEMENT OF E. M. FS. IN THE SEGMENTS OF A QUADRIPOLAR ARMATURE.

so that the circuits through the armature segments are complete ; then the following conditions will hold :

	E. M. F. Volts.	Drop. Volts.	Current. Amperes.	Wasted Power. Watts.
A =	104	1.5	150	225
B =	103	0.5	50	25
C =	102	−0.5	−50	25
D =	101	−1.5	−150	225
			0	500

An inspection of these values shows that a difference of three volts between the E. M. Fs. of the four segments, produces a reversal of current through C and D, at no load, with a useless expenditure of 500 watts. Consequently, between no load and full load, there will be a change from an expenditure of power with reversal of current in the weaker segments, to an excessive drop and expenditure of power without reversal of current.

159. Although this difficulty, arising from the unbalanced magnetic position of the armature, does not, in practice, give rise to any serious inconvenience, when mechanical construction is carefully attended to, yet windings have been devised by which it may be altogether avoided. For example, if all the turns be so connected that their E. M. Fs. are placed in series, then a single pair of brushes will be capable of carrying the current from the entire armature, which will only be divided into two circuits; or, the segments may be so interconnected that turns in distant segments may be connected in series so as to obtain a more general average in the total E. M. F. Such windings are always more or less complex, and the reader is referred to special treatises on this subject for fuller details.

160. The formula for determining the E. M. F. of a multipolar Gramme generator armature is,

$E = \Phi nw$ C. G. S. units, where Φ, is the useful flux in webers, or the flux entering the armature through each pole, n, the number of revolutions per second of the armature, and w, the number of turns on the surface of the armature counted once around. If, however, the armature be series connected, so that instead of having p, circuits through it between the brushes, where p, is the number of poles, there are only two circuits, then the E. M. F. will be $E = \frac{p}{2} \Phi nw$, while if, as in some alternators, the circuit between the brushes be a single one, the mean E. M. F. of the armature will be $p\Phi nw$.

161. Fig. 129 represents the magnetic circuits of an octopolar generator, the dimensions being marked in inches and in centimetres. The field frame is of cast steel, and the armature core is formed of soft iron discs. Let us assume that there are 768 turns of conductor in the armature winding, and that the speed of rotation is 172 revolutions per minute, or 2.867 per second.

Assuming an intensity of 9,500 gausses in the armature, it may be required to determine the E. M. F. of the machine.

The cross-section of the armature is $31.1 \times 13 = 404.3$ sq. cms., but allowing a reduction factor of 0.92 for the insulating material between the discs, the cross-section of iron is 372 sq. cms. The total flux passing through the cross-section of the armature will, therefore, be $372 \times 9,500 = 3,534,000$ webers.

The useful flux through each pole will be twice this amount, or 7,068,000 webers, so that the E. M. F. of the generator will be :

$E = \Phi nw = 7,068,000 \times 2.867 \times 768 = 1.557 \times 10^{10} = 155.7$ volts.

This will be the E. M. F. of the generator, provided all the

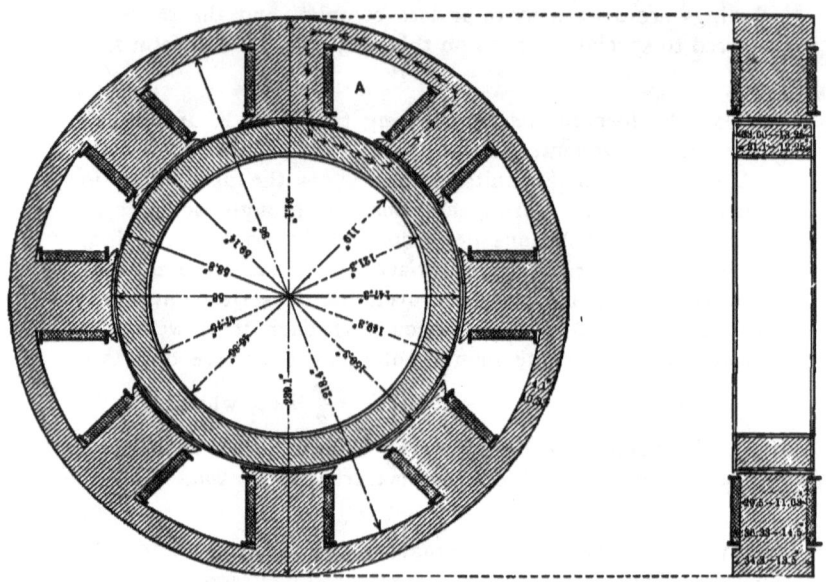

FIG. 129.—GRAMME-RING OCTOPOLAR GENERATOR.

armature segments are connected in parallel, as shown in Fig. 115. If, however, the armature winding be so connected that only a single pair of brushes and a single pair of circuits exist through the armature, the E. M. F. would be 4 times as great, while if the armature could be connected in a single series, the E. M. F. would be 8 times as great.

162. In order to determine the M. M. F. necessary to drive this flux through the armature we proceed as follows: viz.,

MULTIPOLAR GRAMME-RING DYNAMOS.   151

We first determine the cross-section, the mean length, and the intensity in each portion of the magnetic circuits. One of the eight magnetic circuits through the armature is represented by the dotted arrows at *A* (Fig. 129). We may assume that the flux through the cores is 7,068,000 × 1.3 = 9,188,400 webers; 1.3, being the approximate *leakage factor* for a machine of this type; in other words, of all the flux passing through the cores $\frac{10}{13}$ × 100 = 76.9 per cent., approximately, may be assumed to pass through the armature, half through each cross-section. Consequently, we have the following distribution:

	Cross-section. Sq. cms.	Flux. Webers.	Intensity. Gausses.	Length Cms.
Field core,	684	9,188,400	13,430	40
Yoke,	354	3,534,000	9,980	76
Armature,	372	3,534,000	9,500	50

The entrefer, or gap, of copper, air and insulation, existing between the iron in the armature and in the pole faces, is 1.5 centimetres in length, while the polar area is 41 cms. × 34 cms., or 1,400 sq. cms. in cross-section. From these data, the reluctance in the magnetic circuit through the armature is

	Length. Cms.	Intensity. Gausses.	Reluctivity.	Cross-section. Sq. cms.	Cross-section carrying armature flux. Sq. cms.	Reluctance. Oersted.
Field core,	40	13,430	0.002	342	263.1	0.000,304
" "	40	13,430	0.002	342	263.1	0.000,304
Yoke,	76	9,980	0.001	354	354.0	0.000,215
Entrefer,	1.5		1.	700		0.002,142
"	1.5		1.	700		0.002,142
Armature,	50	9,500	0.0008	372	372	0.000,107,5
						0.005,214.5

The M. M. F. required to drive a total flux of 3,534,000 webers through this circuit will be

3,534,000 × 0.005,214,5 = { 18,430 gilberts.
14,665 ampere-turns.
7,333 ampere-turns on each spool.

With 600 turns on each spool, the current would be 12.22 amperes.

CHAPTER XIV.

DRUM ARMATURES.

163. The *drum armature* was first introduced into electrical engineering by Siemens, in the shape of the shuttle armature, and was modified by Hefner-Alteneck in 1873. The drum armature was subsequently modified in this country by the introduction of a laminated iron armature core, consisting of discs of soft iron, called *core discs*, provided with radial teeth or projections. This armature core, when assembled, had

FIG. 130.—TOOTHED-CORE ARMATURE IN VARIOUS STAGES OF CONSTRUCTION.

space provided between the teeth for the reception of the armature loops on its surface, a completed armature showing, when wound, alternate spaces of iron and insulated wire, and formed what is called a *toothed-core armature*. Next followed the *smooth-core drum armature*, that is, a drum armature composed of similar core discs in which the teeth were absent, so that the completed armature had its external surface completely covered with loops of insulated wire. Fig. 130 shows a common type of toothed-core armature in various stages of construction. The laminated iron core is shown at *A*, as assembled on the armature-shaft ready to receive its winding of conducting loops in the spaces between the radially projecting teeth. At *B*, is shown the same core provided with windings of insulated wire. At *C*, is shown a completed armature. The detailed construction of a laminated armature core is illustrated in Fig. 131, which shows a portion of a drum armature core already assembled by the aid of large bolts passing

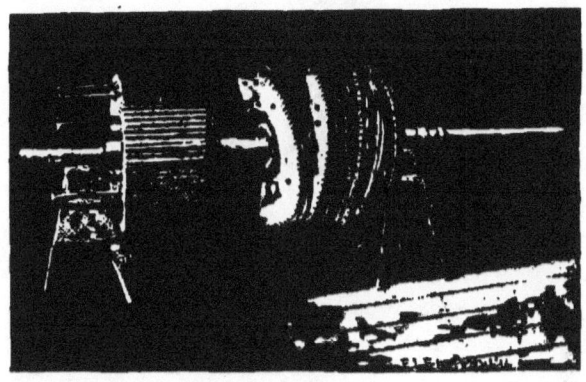

FIG. 131.—TOOTHED-CORE ARMATURE IN PROCESS OF ASSEMBLING.

through holes in the core-discs. On the right are other core-discs ready to be placed in position on the shaft.

164. Fig. 132 shows a laminated armature body of the smooth-core type. Here the separate core-discs are formed

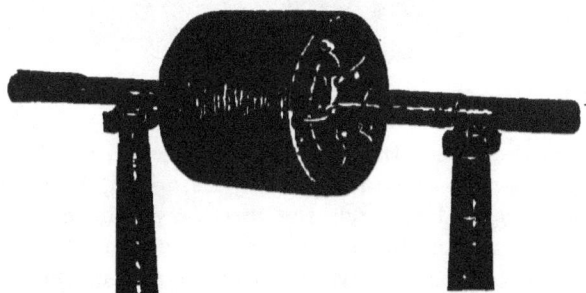

FIG. 132.—SMOOTH-CORE ARMATURE BODY.

of sheet iron rings assembled on the armature shaft as shown. These discs, after being assembled, are pressed together hydraulically. The end rings are heavy bronze spiders, held together internally by six bolts shown in the figure. When the armature body is subjected to compression, these bolts are tightened on the spiders, which are firmly keyed to the shaft, so that on release of the hydraulic pressure, the lami-

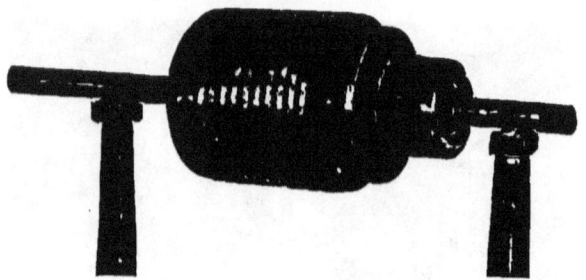

FIG. 133.—COMPLETED ARMATURE, SMOOTH-CORE TYPE.

nated iron core forms one piece mechanically. Fig. 133 shows the same armature completely wound.

165. In the drum armature, the conducting wire is entirely confined to the outer surface, and does not pass through the

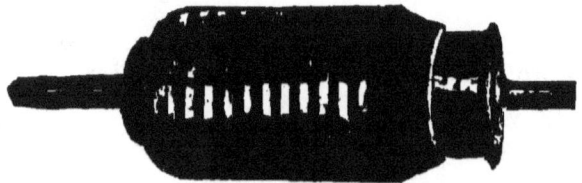

FIG. 134.—TYPICAL FORM OF SMALL SIZE DRUM ARMATURE.

interior of the core. In this respect, therefore, it differs from the Gramme-ring armature, already described, in which the winding is carried through the interior of the core, lying, therefore, partly on the interior and partly on the exterior. The armature core, or body, of a Gramme-ring machine differs markedly in appearance from the armature body of a drum machine, when both are in small sizes, since then the drum core is practically solid, having no hollow space, so that it would be impossible to wind it after the Gramme method. Such a drum-wound armature is shown in Fig. 134. When, however, the drum armature is increased in size, so as to be employed in multipolar fields, the form of the core or body passes from a solid cylinder to that of an open cylinder or ring, as is shown in Figs. 132 and 135, so that it would be possible to place a conducting wire on such a core either after the drum or Gramme type of winding. The tendency, however, in modern electrical engineering is, perhaps, toward the production of drum-wound rather than Gramme-wound generators.

FIG. 135.—LARGE DRUM ARMATURE FOR MULTIPOLAR FIELD.

This tendency has arisen, probably more from mechanical and commercial reasons than from any inherent electrical objections to armatures of the Gramme-ring type.

166. The windings of drum armatures are numerous and complicated in detail, but all may be embraced under two typical classes; namely, *lap-winding* and *wave-winding*. In lap-winding, the wire is arranged upon the surface of the armature in conducting loops, the successive loops overlapping each other, hence the term; while in wave-winding, the conducting wire makes successive passages across the surface of the armature, while being advanced around its periphery in the same direction.

167. Lap-winding is applicable particularly to bipolar armatures, while wave-winding is applicable only to multipolar armatures.

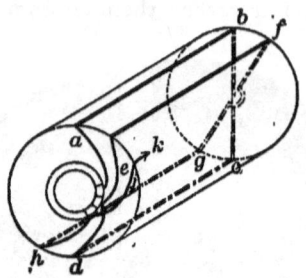

FIG. 136.—SIMPLE BIPOLAR DRUM-WINDING.

The simplest form of lap-winding is shown in Fig. 136, where the separate paths taken by the turns a, b, c, d, and e, f, g, h, across the outside of the bipolar armature core, and their connections to the commutator, are represented as shown. If the

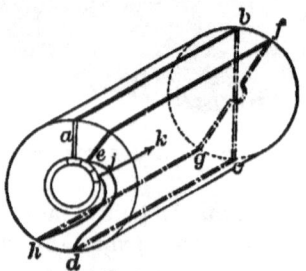

FIG. 137.—SIMPLE BIPOLAR DRUM-WINDING WITH LEAD IN COMMUTATOR CONNECTIONS.

entire winding of the armature be completed, it is evident that any attempt to represent the winding graphically by the method adopted in this figure would lead to great complexity. For this reason it is customary to represent the armature surface as unrolled, or developed upon the plane of the paper, as shown in Fig. 138. For example, the winding already shown in Fig. 136 becomes on this development represented as in Fig. 138. Here it is clear that each loop overlaps its prede-

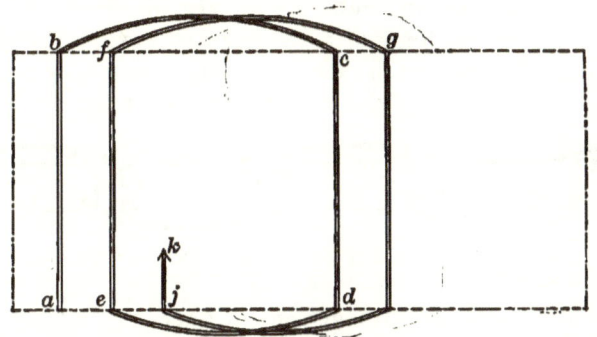

FIG. 138.—DEVELOPMENT OF LAP-WINDING IN FIGS. 136 AND 137.

cessor, and, consequently, it is evident that the simplest form of drum-winding is a lap-winding.

Fig. 137 represents the same winding as Fig. 136, except

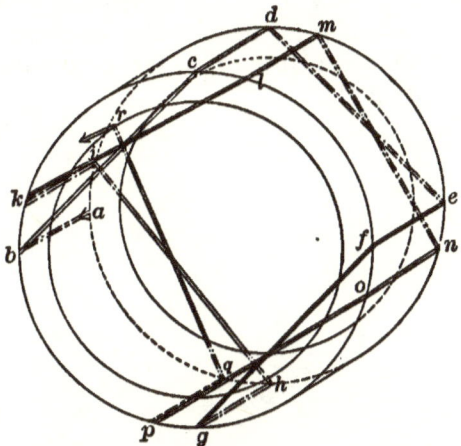

FIG. 139.—QUADRIPOLAR WAVE-WINDING.

that the connections with the commutator are given a lead of 90 degrees, requiring a correspondingly altered position of the brushes of the machine.

Fig. 139 represents a number of conductors, *ab, cd, ef, gh,* etc., wound on the external surface of a drum core in the winding of the wave type. Here it will be seen that the conducting wire, after crossing over from one side of the armature core to the other, advances progressively over its surface in the form of a rectangular wave. The corresponding development is shown in Fig. 140. The winding shown is applicable only to multipolar fields; for, an inspec-

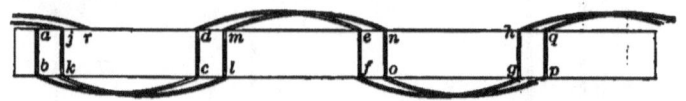

FIG. 140.—DEVELOPMENT OF QUADRIPOLAR WAVE-WINDING.

tion of this particular arrangement of wave-winding will show that when conducting wires *ab* and *ef* are passing north poles, the conducting wires *cd* and *gh*, are passing south poles, and the direction of the induced E. M. F. while opposite in successive conductors, as regards the separate conductors *ab*, *cd*, *ef*, and *gh*, is, nevertheless, unidirectional, so far as the entire circuit *a, b, c, d, e, f, g, h, i, j, k*, is concerned. In the same manner a wave-winding for an octopolar machine is required to be spaced in accordance with the successive distances between alternate poles.

CHAPTER XV.

ARMATURE JOURNAL BEARINGS.

168. Even in the best designed types of electro-dynamic machinery, there are certain losses of electric energy which necessarily occur in the operation of the machine. These losses may be grouped under the general head of frictions, and include mechanical, electric, and magnetic frictions. Since in well-designed types of large machines the commercial efficiency may be as high as 95 per cent., it is evident that the

FIG. 141.—SIGHT-FEED LUBRICATED BEARING.

losses from all these causes combined can be kept within a small percentage of the total output.

169. This high efficiency, however, can only be obtained in the case of large machines. In those of smaller output, the proportion of the losses may be much greater. It is, therefore, advisable to examine the causes of these various losses, their variation with the output of a machine, and the means by which they are commercially reduced.

Considering first the mechanical losses : these may exist as friction in the bearings of the moving parts of the generator, friction arising from the pressure of the brushes on the commutator, or contact parts, and friction from air churning.

The journal bearings are lubricated either by sight-feed oiling, or self-oiling devices. In *sight-feed oiling devices*, a glass oil cup, filled from time to time with oil, allows oil to drop slowly on the journal bearings, but requires to have its outlets opened by hand, when the machine commences to run, and also to be stopped when the machine stops.

Fig. 141 represents an end view and longitudinal section of

FIG. 142.—DETAIL OF SELF-OILING BEARING.

such a bearing. The oil cup $C C$, is provided with a head H, by the rotation of which an outlet in the base is adjusted. The oil descends by gravity to the shaft $S S$, where, by the movement of the shaft, it is mechanically carried through spiral grooves on the inner surface of the babbitt-metal sleeve $B B$, passing finally, from the ends on the bearing, into the pans PP, whence it is drawn off at intervals and filtered.

The upper pan, P, is intended to catch any overflow of oil that may occur during the process of filling. The box $X X$, enclosing the babbitt-metal sleeve, is capable of rotation within small limits, about a vertical axis, upon the spherical surfaces $Z Z$. This play admits of the true alignment of the bearings to the shaft $S S$. As soon as the shaft has been introduced and becomes self-aligned, any further undue play in the bearing is prevented by tightening the nuts $N\ N$.

170. Sight-feed lubricating bearings necessitate, as already observed, the opening and closing of the oil cup at the starting and stopping of the machine. They have been, consequently, almost entirely replaced by *self-oiling bearings*, which require no such attention; here the oil is automatically fed to the revolving shaft by its rotation. A self-oiling bearing of this description is represented in Fig. 142. The oil is supplied to the bearing into the oil well $O\ O$, through a screw hole h.

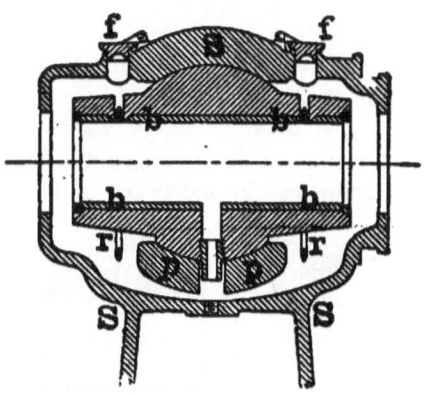

FIG. 143.—LONGITUDINAL SECTION OF SELF-OILING BEARING.

During the rotation of the shaft $S\ S$, two rings $r,\ r$, which rest upon the upper surface of the shaft, and dip into the oil within the well, are set in rotation, and carry oil on the surface of the shaft, where it is spread over the bearing along suitable grooves in the babbitt-metal sleeve, as in the previous case. Grooves are made in the upper surface of the babbitt-metal sleeve for the reception of the rings, and the rings are prevented from leaving the grooves by the screw clips $m,\ m$. The rings are carried around by the friction caused by their weight as they rest on the shaft, and, therefore, do not necessarily rotate as rapidly as the surface of the shaft. The babbitt-metal sleeve, which holds the shaft, is contained in a cylindrical box with a spherical bolt B, at its centre. A pin or projection *p*, at the bottom of this box, engages in a hole in the frame work, thus preventing the box from rotating with the shaft, but enabling the shaft to align itself freely in the sleeves. Nuts *n*, of which only one is seen, clamp the box *B*, in position.

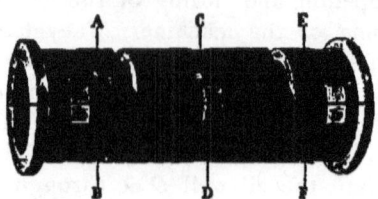

FIG. 144.—SLEEVE OF BABBITT METAL IN JOURNAL BEARING.

A draw-off cock is provided at *d*, for withdrawing the oil from the well at suitable intervals.

171. Fig. 143 represents a longitudinal cross-section of a similar bearing employed in machines of larger size. Here

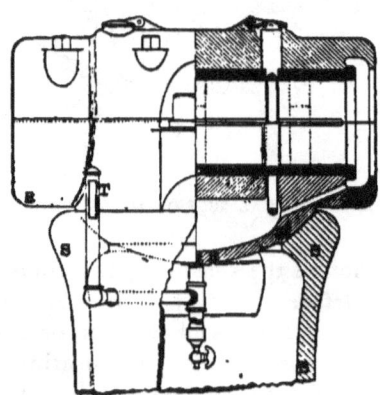

FIG. 145.—DETAILS OF LARGE SELF-OILING JOURNAL BEARING.

oil is fed through two openings *f*, *f*, and accumulates in the lower part of the hollow cast-iron support *S S*. The rings *r*, *r*, by their revolution upon the shaft, carry the oil into the babbitt-metal sleeve *b b b b*, as before. The shaft is supported upon the bracket *p p*, which forms part of the pedestal or support *S S*, and is hollowed spherically so as to permit of the alignment of the babbitt-metal sleeve and its box. Fig. 144 shows a general view of the babbitt-metal sleeve with grooves for the reception of the oil rings, and with lugs $L, L, L,$ for assisting in the aligning. Fig. 145 represents partly in elevation, and partly in longitudinal section, a similar bearing sometimes employed in still larger machines, differing from the last described only in details of construction. The weight of the shaft is taken directly upon the lower half of the bearing $B B B$, which has its lower surface bowl-shaped, and fitting into a pedestal or support SS, in such a manner that the bearing can be readily aligned and finally tightly secured in place by suitable bolts. The gauge glass T, enables the level of the oil in the bearing to be clearly discerned.

172. The amount of energy expended as friction in journal bearings varies with the weight supported on the bearing, the accuracy of the workmanship, the correctness of the alignment, the nature of the lubricating material, the character of the surfaces in contact, the speed of rotation and the diameter of the shaft.

Other things being equal, the energy expended is proportional to the diameter of the journal in the bearing. In order to keep the friction as low as possible, the diameter of the journal is always kept as low as is consistent with ample mechanical strength.

The power expended in brush friction depends upon the number of brushes and the pressure with which they bear upon the commutator. It also increases with the diameter of the commutator and with the speed of rotation of the armature. This waste of energy is often an appreciable fraction of the total waste in a small machine, but is usually quite negligible in a large one.

CHAPTER XVI.

EDDY CURRENTS.

173. During the rotation of the armature of a dynamo-electric machine through the flux produced by its field magnets, electromotive forces are not only generated in the conducting loops on the armature, by the successive filling and emptying of these loops with flux, but they are also generated in all masses of metal revolving through the flux; in other words, the iron in the armature core and the copper of the conductors will also be the seat of E. M. Fs. Though these E. M. Fs. may be locally very small, yet, since the resistances of their circuits are generally exceedingly small, the strength of the currents set up may be very considerable.

Such currents are generally known as *eddy currents*. They are necessarily alternating in character, their frequency depending upon the speed of revolution and upon the number of poles.

Not only is the energy expended in eddy currents lost to the external circuit, since these currents cannot be made to contribute to the output, but such currents also unduly limit the output of the armature, by raising its temperature, independently of the increase of temperature due to the passage of the useful armature current through the conducting loops. Losses of energy due to eddy current are of the type $I^2 R$ (in watts), I, being the strength of the local current in amperes, and R, the resistance of the local circuit in ohms.

174. It is evident that a dynamo machine can never be designed so as to be entirely free from eddy currents; for, conducting loops must be placed on the armature, and, moreover, in nearly all the types of practical dynamo machines, iron armature cores are employed.

All that can be done is to reduce these losses as far as is commercially practicable. In the case of the iron core, for example, the advantage arising from its use; namely, the decrease in the reluctance of the magnetic circuit, can be retained, provided the material of the core is laminated, *i. e.*, made continuous in the direction of the magnetic flux paths, and discontinuous at right angles to this direction.

175. If a piece of metal be revolved in a magnetic field, it will enclose magnetic flux. A distribution of E. M. Fs. will be established in it according to the rate at which the enclosure takes place, and depending upon the shape of the piece. These E. M. Fs. will produce eddy currents in the moving metal. The rate of expending work in eddy currents will be, for a given flux intensity in the metal, in direct proportion to the conductivity of the material. A piece of revolving copper will have much more work expended in it by eddy currents than a piece of lead or German silver. If, however, we divide the mass of metal into a number of segments or smaller portions, the total E. M. F. at any instant will be divided into a number of parts, one in each segment, and the resistance of each segment to its E. M. F. will be much greater than the ratio of the resistance of the entire mass to the total E. M. F. in such mass. The energy wasted in the mass will therefore be reduced. For this reason, the iron core of the armature is divided into sheets or laminæ, in such a manner that the sheets afford a continuous path to the magnetic flux, but no circuit is provided for eddy currents across the sheets. The magnetic flux is conducted through the entire length of the sheet, but the circuits of the eddy currents are all in the cross-sections of the sheet. The division of the armature core does not, therefore, increase the magnetic resistance, or reluctance of the armature, but enormously increases its resistance to eddy currents.

176. Fig. 146 represents at D, an armature core of solid iron capable of being revolved in a quadripolar field N^1, S^1, N^2, S^2, the arrows indicating the general directions of the flux paths. The cross-section of the armature is shown at A, and the arrows represents diagrammatically the distribution of the eddy currents set up in the solid mass of iron during the rotation of the armature. At B, the cross-section is represented with laminations, parallel to the axis of the armature, as, for example, when the armature core is composed of a spiral winding of sheet-iron ribbon. Here the eddy currents are limited to the cross-section of each band or lamina. The magnetic flux, however, has to penetrate all the discontinuities between the bands, in order to penetrate to the deepest layer, unless the flux be admitted to the armature on its sides, as shown in Fig. 8.

At C, the armature is laminated in planes perpendicular to the axis, or is built up of sheet discs. Here the eddy currents are confined, as in the last instance, to the section of each disc, but the flux passes directly along each sheet.

While, therefore, the methods of construction indicated at

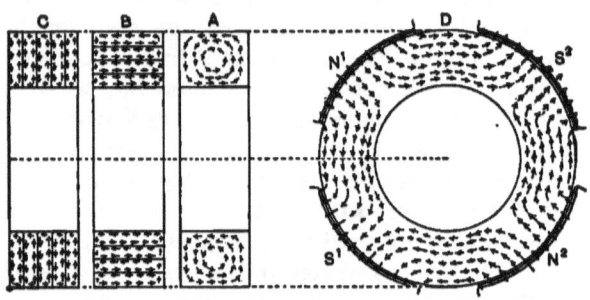

FIG. 146.—DIAGRAM ILLUSTRATING EFFECTS UPON EDDY CURRENTS OF LAMINATING ARMATURE CORES.

B and C, are equally favorable to the suppression of eddy currents, B, tends to increase the reluctance of the armature, and to magnetically saturate the outer layers of the core, with a corresponding sparsity of flux in the inner layers, except when the field poles cover the sides of the armature.

177. Taking a single lamina of the armature core, it is clear that if the intensity in the core is, say, 12 kilogausses, each square centimetre of cross-section in the lamina is linked with 12 kilowebers, first in one direction and then in the opposite direction, as the armature moves from one pole to the next. The value of the E. M. F. round the cross-section of the lamina, considered as a loop, depends upon the speed with which the linkage takes place, and, therefore, on the intensity $\mathfrak{B}$ the speed of rotation and the number of poles. The average E. M. F. in a lamina, rotating at a given speed through a quadripolar field of intensity $\mathfrak{B} = 12{,}000$, would be four times as great as when passing through a bipolar field of intensity $\mathfrak{B} = 6{,}000$. The rate at which an E. M. F. of e volts expends energy in a resistance of r ohms, being $\frac{e^2}{r}$ watts, the average wasteful activity in eddy currents depends upon the square of the speed of magnetic reversal in the core, and also upon the square of the intensity. If, then, we double the speed of revolution in an armature core, we quadruple the eddy current waste of power. The higher the intensity of magnetic flux in the armature, and the more rapid the reversal, the more important becomes the careful lamination of the armature, but the eddy-current-loss in armature cores is usually very small when the plates have a thickness not exceeding 0.02″.

Moreover, when powerful eddy currents are present, the M. M. F. they establish has such a direction as opposes the development of magnetic flux by the field, so that the existence of powerful eddy currents in an armature core tends to shield the interior of the core, or its laminæ, from magnetic flux, thereby reducing the effective cross-section of the armature, or increasing its apparent reluctance. This effect is usually small in revolving armatures at ordinary speeds of rotation, but becomes appreciable when the frequency of reversal is high and the degree of lamination insufficient.

178. It used to be the universal practice to separate adjacent sheets of iron by thin sheets of paper, when assembling the cores of armatures, so as to ensure the complete insulation of the separate laminæ. This introduction of paper into the core had the disadvantage of reducing the effective permeance of the armature core, or in other words, of increasing the flux density in the iron. It has been ascertained experimentally, however, in recent times, that the paper could usually be dispensed with, as the superficial layer of oxide on the iron sheets formed a layer of sufficient resistance to effectually insulate the laminæ against the feeble E. M. Fs. in the eddy current circuits.

179. As we have seen, eddy currents are not limited to the iron core of an armature, but are also set up in the conductors wound on the armature.

In this case, eddy currents are set up in their substance by revolution under the poles, but the conditions differ slightly in detail. A Grammè-ring armature, for example, has no eddy currents set up in the conductors except upon the outer surface of the armature, since the flux passes through the wire at the outer surface and not through the wire on the inner surface. Similarly, a drum armature has no eddy currents set up in the wire upon the ends of the drum, if we may neglect such leakage flux as may pass through the ends of the core. Again, the amount of eddy-current-loss will depend upon the distribution of the magnetic flux over the surface of the armature. If the flux entering the armature terminates sharply at the edge of the pole-pieces, so that the wire suddenly enters or suddenly leaves a powerful magnetic field in the air-gap, the rate of change of the flux enclosed in the substance of the wire will rapidly vary, inducing a brief, but powerful, E. M. F. in its substance, and the total expenditure of energy by eddy currents will be considerably greater than if the gradient of magnetic intensity in the neighborhood of the polar edges is less abrupt, and the E. M. F. smaller in amount but more prolonged.

180. The eddy-current-loss for a given size of machine is apt to be considerably greater with low pressure than with high pressure armatures, since the former require few massive copper conductors, while the latter require many, separately insulated, conductors. The plan is, therefore, frequently adopted of winding low-pressure smooth-core armatures with multiple conductors, each main conductor being composed of a cable of separately insulated wires. Even when this is done, an additional precaution is necessary, namely, to transpose the conductors or twist them through 180 degrees, halfway across the armature surface, in order to prevent any pair of wires from acting as a loop for the generation of the E. M. Fs. This is illustrated diagrammatically in Fig. 147, where the multiple conductor CC^1, consisting of five insulated wires, laid over the surface of the armature core $A\ A\ A\ A$, is reversed in the centre, so that the advancing wire at one end becomes the receding wire at the other, and vice-versa.

It is sometimes found that the insertion of a sheet iron cylinder of the form outlined in Fig. 148, closely fitted into the polar bore, and forming a tube within which the armature revolves, greatly diminishes the waste of energy in eddy currents. This is for the reason that the edges of the pole-pieces are removed, and the flux through the entrefer gradually varies between zero and full intensity as we advance round the field. The effective area of the polar surfaces is for the same reason increased. The objection to the introduction of such a cylinder lies in the magnetic leakage it introduces; for,

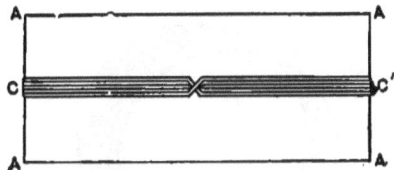

FIG. 147.—DIAGRAM INDICATING THE TRANSPOSITION OF MULTIPLE ARMATURE CONDUCTOR.

if S, be the cross-section of the soft iron sheet in square centimetres, the flux it will carry, direct from pole to pole, will be roughly 20,000 S, webers, and this flux has to be provided for through the magnetic circuit of the field frame in addition to other leakage and the useful flux through the armature.

181. When the armature conductors are buried beneath the surface of the iron, as, for example, when they run in the deep grooves of toothed-core armatures, practically no eddy currents are produced in them, for the reason that the space they occupy is almost free from the flux established by the field. A toothed-core armature may, therefore, be considered as an armature in which the eddy currents are confined to the iron laminæ of the core. This feature constitutes one of the advantages of toothed-core armatures.

182. Besides the eddy currents set up in the armature, and in the conducting masses of the metal on the armature, they also occur in the edges of the pole-pieces of the field magnets, both in the case of dynamos and motors. The strength of these eddy currents is greater in the pole which is approached by a generator armature, and in that which is receded from by a motor armature, as is evidenced by the fact often observed, that, although both polar edges become warm during the action of the machine, one edge becomes warmer than the other. The reason for this difference will be considered later.

183. The tendency to the development of eddy currents in pole-pieces is increased when the armature is changed from a smooth core to one of the toothed-core type. The reasons for this are twofold; in the first place, in the toothed-core arma-

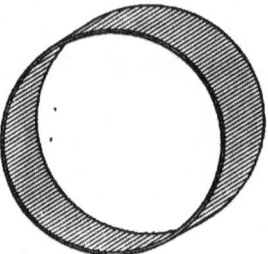

FIG. 148.—IRON CYLINDER OR BUSHING FOR FIELD CORE.

ture the armature is brought nearer to the pole face, so that all magnetic disturbances in the armature are more likely to set up corresponding disturbances in the poles; in the second place, because the revolving teeth set up waves of magnetization in the polar surfaces, thus giving rise to the development of eddy currents. Consequently, the change from a smooth-core to a toothed-core armature suppresses the eddy currents in the wire on the armature, but creates, or tends to create, eddy currents in the pole-pieces.

184. In some types of machines the pole-pieces are grooved or slotted, so as to check the development of eddy currents, just as the armature is in effect grooved or slotted by the use of laminated cores. An example of this is seen in Fig. 14. In fact, some field magnets are constructed of a frame of cast iron or cast steel, with receptacles within which are placed the pole-pieces formed of a number of iron plates bolted together, the laminations extending in the same direction as those in the armature beneath.

It will be evident that there can be no tendency to set up eddy currents in the solid cores of the field magnets excited by steady, continuous currents. Consequently, no advantage is derived from a lamination of field magnet cores at distances beyond the influence of magnetic changes produced by the teeth or conductors on the revolving armature.

CHAPTER XVII.

MAGNETIC HYSTERESIS.

185. Besides the losses in the iron masses of a dynamo due to eddy currents, there are others in the same masses due to *magnetic friction* or *hysteresis.* These latter losses, like the others, are dissipated as heat.

The losses due to hysteresis occur in nearly all forms of dynamo-electric machinery. In continuous-current generators these losses are practically limited to the armature; in some forms of alternating-current machines, they exist both in the armature and field, and are especially present in alternating-current transformers. It becomes, therefore, a matter of no little importance to thoroughly understand the nature of this source of loss.

186. A certain amount of energy has to be expended in order to magnetize a bar of iron. This energy resides in the magnetic flux passing through the magnetic circuit of the bar. The energy is transferred from the magnetizing circuit by the production of a C. E. M. F. in the magnetizing coil, and this C. E. M. F. e, (usually very small), multiplied by the magnetizing current strength I, at that moment, gives as the product eI, the activity expended in producing the magnetic field. As soon as the full magnetic flux is established, the C. E. M. F. ceases, being dependent upon the rate of change of flux enclosed, so that no more energy is expended in the iron, and the current only expends energy as i^2r, in heating the magnetizing coil. When the magnetizing current is interrupted, say by short circuiting the source of E. M. F. in the circuit, the magnetism in the bar tends to disappear, and, as the magnetic flux diminishes, an E. M. F. is set up in the coil, tending to prolong the action of the waning magnetizing current. In other words, the E. M. F. set up in a circuit by the waning magnetic flux is such as will tend to do work on the current, with an activity of the type ei watts, and, in this manner, restore to the circuit the energy expended in the magnetization. Were all the energy in this case returned to the circuit, there would be no loss by hysteresis. As a matter of fact, however, while practically all such energy would be returned to the circuit, if the coil magnetized air, wood, glass, etc., yet, when the coil magnetizes iron, although a greater magnetic flux is obtained, yet some of the energy is not restored, but is expended in the iron as heat.

187. It is now generally believed that each of the molecules in a mass of iron is naturally and permanently magnetized, so that each molecule may, therefore, be regarded as a molecular compass needle. In the ordinary unmagnetized or neutral condition of iron, these separate molecular magnets possess no definite alignment, and, consequently, neutralize one another's influence by forming closed loops or chains. When the iron becomes magnetized by subjection to a magnetizing force, these loops break up and become polarized or aligned, all pouring their magnetic flux in the same direction; $i\ e.$, parallel to the magnetic axis. When the magnetizing force is removed, the molecular magnets tend to resume their old positions; and, if they did resume exactly their old positions, the magnetism in the iron would entirely disappear on the removal of the magnetizing force, and all the magnetic energy would be restored to the circuit. In point of fact, however, they do not exactly resume old positions, but take new intermediate positions, by virtue of which a certain amount of *residual magnetism* is left in the bar.

188. When now the magnetizing force is reversed, by reversing the current through the magnetizing coil, the molecules are forced around, and breaking suddenly from their positions, fall into new positions, either with oscillations, or with a frictional resistance to the motion, that dissipates energy as heat. The energy thus lost by molecular vibration or molecular friction cannot be returned to the circuit. Consequently, a loss of energy occurs in the circuit supplying the reversing magnetizing force, at each reversal of magnetism in the magnetized iron, since the opposing E. M. Fs. developed

in the coil during magnetization and demagnetization are not equal, and the energy so lost results in an increase in temperature of the iron. By *hysteresis*, (his-ter-ee'-sis), is meant that property of iron, or other magnetic metal, whereby it tends to resist changes in its magnetization when subjected to changes in magnetizing force. That is to say, when a mass of iron is successively magnetized and demagnetized, or passes through *cycles of magnetization*, the magnetic intensity in the mass *lags behind* the impressed magnetizing force. The word hysteresis take its origin from this fact, since it is derived from a Greek word meaning to lag behind. This phenomenon is called hysteresis, and the loss of energy due to this cause is called *hysteretic loss*, or loss of energy by hysteresis.

189. When iron undergoes successive magnetic reversals, the amount of hysteretic loss is found to depend upon the maximum magnetic intensity in the iron at each cycle; that is to say, upon the maximum value of $\mathcal{B}$. As $\mathcal{B}$, increases, the amount of work that has to be expended in reversing the magnetization increases, and if we double the value of $\mathcal{B}$, we practically treble the amount of work that has to be expended. It was first pointed out by Steinmetz, as a consequence of this relation, that the hysteretic loss varied as the 1.6th power of $\mathcal{B}$, or as $\mathcal{B}^{1.6}$, the formula for the amount of activity expended in one cubic centimetre of magnetic metal being $P = n \eta \mathcal{B}^{1.6}$ watts. Since the same loss of energy occurs in a cubic centimetre during each cycle, the more rapidly the cycles recur, the greater will be the wasteful activity, and n, in the above formula, expresses the number of complete cycles through which the iron is carried per second. The coefficient η, is the *hysteresis coefficient* for the metal considered, and has to be determined experimentally. It may be regarded as the activity in watts which would be expended in one cubic centimetre of the metal when magnetized and demagnetized to a flux density of one gauss at one complete cycle or double reversal per second. The following table gives the values of this coefficient, and also the amount of hysteretic loss produced in a cubic centimetre, and in a pound, of ordinary good commercial sheet iron at various frequencies and intensities.

Table Showing the Hysteritic Activity in Good, Soft Sheet Iron or Steel Undergoing One Complete Magnetic Cycle per Second, in Watts per Cubic Centimetre, Watts per Cubic Inch, and Watts per Pound, for Various Magnetic Intensities in Gausses and in Webers per Square Inch.

Webers, per sq. in....	6,452	12,900	19,360	25,810	32,260	38,710	45,160	51,620	58,060
Gausses [$\mathfrak{B}$].	1,000	2,000	3,000	4,000	5,000	6,000	7,000	8,000	9,000
Watts, per cc.	1.58×10^{-5}	4.78×10^{-5}	9.15×10^{-5}	1.45×10^{-4}	2.07×10^{-4}	2.78×10^{-4}	3.55×10^{-4}	4.40×10^{-4}	5.31×10^{-4}
Watts, per cubic in...	2.59×10^{-4}	7.84×10^{-4}	1.50×10^{-3}	2.38×10^{-3}	3.40×10^{-3}	4.55×10^{-3}	5.82×10^{-3}	7.20×10^{-3}	8.69×10^{-3}
Watts, per lb.	9.17×10^{-3}	2.78×10^{-2}	5.32×10^{-2}	8.43×10^{-2}	1.21×10^{-1}	1.62×10^{-1}	2.06×10^{-1}	2.56×10^{-1}	3.09×10^{-1}

Webers, per sq. in....	64,520	70,960	77,420	83,860	90,320	96,770	103,200	109,700	116,100
Gausses [$\mathfrak{B}$].	10,000	11,000	12,000	13,000	14,000	15,000	16,000	17,000	18,000
Watts, per cc.	6.28×10^{-4}	7.31×10^{-4}	8.40×10^{-4}	9.55×10^{-4}	1.08×10^{-3}	1.20×10^{-3}	1.33×10^{-3}	1.47×10^{-3}	1.61×10^{-3}
Watts, per cubic in...	1.03×10^{-2}	1.20×10^{-2}	1.38×10^{-2}	1.57×10^{-2}	1.76×10^{-2}	1.97×10^{-2}	2.18×10^{-2}	2.41×10^{-2}	2.64×10^{-2}
Watts, per lb.	3.65×10^{-1}	4.25×10^{-1}	4.89×10^{-1}	5.56×10^{-1}	6.26×10^{-1}	6.79×10^{-1}	7.75×10^{-1}	8.53×10^{-1}	9.35×10^{-1}

190. As an example of the hysteretic activity, we may consider a pound of iron subjected to a periodic alternating flux density of ten kilogausses, with a frequency of 25 cycles-per second. From the preceding table, it is seen that at 10 kilogausses the hysteretic activity is 0.0365 watts-per-pound, at a frequency of one cycle per second. At 25 cycles per second this would be $25 \times 0.0365 = 0.9125$ watt $= 0.9125$ joule-per-second $= 0.6735$ foot-pound per second. Consequently the hysteretic activity might be represented by lifting the pound at the rate of 0.6735 foot per second against gravitational force. If, therefore, all the iron in an armature core be subjected to an intensity of ten kilogausses, and rotates 25 times per second in a bipolar field, 12.5 times per second in a quadripolar field, or 6.25 times per second, in an octopolar field, hysteretic activity is being expended at a rate which is probably represented by the activity of raising the whole armature core about eight inches per second.

It is to be observed that the table represents average samples of good commercial iron, and by no means the best quality of iron obtainable.

191. As an example of the application of this table, suppose that it is required to estimate the power expended in hysteresis during the rotation of the armature of the octopolar generator represented in Fig. 129, the weight of iron in the armature being 2,700 lbs.

At the maximum intensity of 9,500 gausses, or 61,290 webers-per-sq. in., the table shows that the hysteretic activity per pound at one cycle per second is about 3.4×10^{-2} watts. In each revolution of the armature there would be eight reversals, or four complete cycles, and at 2.867 revolutions per second, the frequency of reversal would be 11.468 cycles per second. The total hysteretic activity is, therefore,

$$P \times 2,700 \times 3.4 \times 10^{-2} \times 11.468 = 1,053 \text{ watts.}$$

This would be the hysteretic activity in the armature when generating 155.7 volts. When generating a lower E.M.F., the flux intensity in the armature would be reduced, and, therefore, the hysteretic activity.

192. Hysteresis, therefore, occurs when a mass of iron undergoes successive magnetizations and demagnetizations, and this is true whether such are caused by the reversal of the magnetizing current, with the mass at rest, or by the reversal of the direction of the mass in a constant magnetic field. Consequently, the revolutions of the armature of a dynamo or motor, occasioning the successive magnetizations and demagnetizations of its core, are accompanied by an hysteretic loss of energy.

The amount of this hysteretic loss increases directly with the volume V, of iron in the armature in c. c., the number n, of revolutions of the armature per second, the hysteretic coefficient η of the iron employed, and the 1.6th power of the maximum magnetic intensity in the iron; for, it is evident that

MAGNETIC HYSTERESIS. 177

in one complete revolution of the armature its direction of magnetization will have undergone two reversals, provided that the field is bipolar. In a multipolar field the number of reversals increases with the number of poles p, and the hysteretic activity becomes $P = \dfrac{V\, n\, \eta\, p\, \mathfrak{B}^{1.6}}{2}$ watts. In the case of a generator, this activity must be supplied by the driving power, and in the case of a motor by the driving current.

193. When a generator armature is at rest in an unmagnetized field, the *torque*; *i. e.*, the twisting moment of the force which must be applied to the armature in order to rotate it, is such as will overcome the friction of the journals and brushes. When, however, the field is excited, so that the armature becomes magnetized, the torque which is necessary to rotate the armature is increased, even when the armature is symmetrically placed in regard to the poles. This extra torque is due to hysteresis. It is sometimes called the *hysteretic torque*, and is equal to

$$\tau = \dfrac{V\, \eta\, p\, \mathfrak{B}^{1.6}}{4\pi}\ \text{megadyne-decimetres.}$$

194. The total useless expenditure, therefore, of power in an armature core is the sum of the hysteretic and eddy current loss. The former increases as the speed of revolution directly, but the latter, as already pointed out, increases as the square of the speed. Consequently, if we have an unwound armature core, and rotate it on its shaft through a field which is at first unexcited, we expend an activity which might be measured, and which would be entirely frictional loss. When the field is excited, we expend activity against mechanical friction, hysteresis and eddy currents combined. By varying the speed of rotation, and observing the rate at which the activity given to the rotating armature increases, it is possible to separate the three descriptions of losses from each other.

195. Although, as we have seen, the hysteretic loss increases with the 1.6th power of the intensity of flux, yet it is stated to have been found experimentally, that when a mass of iron, such as an armature, is rotated in a sufficiently powerful magnetic field, the hysteretic loss entirely disappears, owing to the supposed rotation of all the elementary molecular magnets about their axes during the rotation of the armature without losing parallelism, and, consequently, without any molecular oscillation and expenditure of magnetic energy as heat. So far as experiments have yet shown, this critical intensity in the iron is above that which ordinary dynamo or motor armatures attain, so that under practical conditions, the 1.6th power of the maximum intensity determines the hysteretic loss.

196. From an examination of the formula expressing the hysteretic activity in the armature, it is evident that the activity might be decreased by a decrease either in the number of poles, the speed of revolution, the flux density, or the hysteretic coefficient. Since, however, in any machine the first three factors are practically fixed, it is important that the remaining factor, or hysteretic coefficient, should be kept as low as is commercially possible. For this reason, whenever the hysteretic loss is a considerable item in the total losses of the generator, care is taken to select the magnetically softest iron commercially available, in which the hysteretic coefficient is a minimum.

197. We have already referred to the increase in temperature of the edges of the field-magnet poles during the operation of a dynamo armature, and have ascribed the cause of such heating to the development of eddy currents locally produced there. It is to be remarked, however, that some of the heat in such cases may usually be ascribed to true hysteretic changes in magnetization.

CHAPTER XVIII.

ARMATURE REACTION AND SPARKING AT COMMUTATORS.

198. In the operation of a dynamo-electric generator, considerable difficulty is frequently experienced from the *sparking* which occurs at the commutator, that is to say, instead of the current being quietly carried off from the armature to the external circuit, a destructive arc, which produces burning, occurs between the ends of the brushes and the commutator segments. The tendency of this sparking, unless promptly checked, is to grow more and more marked from the mechanical irregularities produced by the pitting and uneven erosion

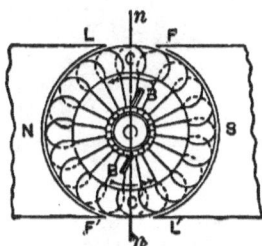

FIG. 149.—GRAMME-RING ARMATURE IN BIPOLAR FIELD ON OPEN CIRCUIT.

of the commutator segments. It becomes, therefore, a matter of considerable practical importance to discuss the causes of sparking at the commutator, and the means which have been proposed, and are in use, to overcome the difficulty.

199. When a Gramme-ring armature, such as that shown in Fig. 149, is rotated on open circuit, in a uniform bipolar field, the brushes, when placed on the commutator, must be kept at diametrically opposite points corresponding to the line *n n*. If applied to the commutator at any other points, sparking will probably occur, although the armature is on open circuit. The reason for this is seen by an examination of the figure, which represents a pair of coils *C, C'*, about to undergo *com-*

mutation; *i. e.*, about to be transferred by the rotation of the armature from one side of the brush to the other, and being short circuited by the brushes, as they bridge over the adjacent segments of the commutator to which their ends are connected. Since the coils C, C', in the position shown, embrace the maximum amount of flux passing through the armature, there will be no E. M. F. induced in them, and, consequently, there will be no current set up during the time of short circuit under the brushes. In other words, the prime condition for non-sparking at the commutator is that the coils shall be short

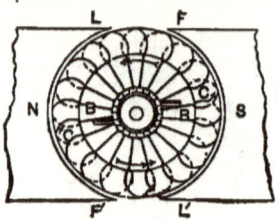

FIG. 150.—GRAMME-RING ARMATURE WITH BRUSHES DISPLACED FROM NEUTRAL LINE.

circuited only at the time, and in the position, where no E. M. Fs. are being generated in them.

200. If the brushes be advanced into a position such as that represented in Fig. 150, so that the *diameter of commutation;* *i. e.*, the diameter of the commutator on which the brushes rest, is shifted from B, B', to a new position, powerful sparking will, probably, be set up, for the reason that in this position the rate of change, in the flux linked with these coils, is considerable, and, consequently, there is a considerable E. M. F. induced in them, so that, when they are short circuited by the brushes, heavy currents tend to be produced in the circuit of these coils according to Ohm's law. If, for example, a bipolar Gramme-ring armature gives passage to a total useful flux of 1 megaweber, and there are 1,000 turns on the armature, and 50 segments in the commutator, then, if the speed of rotation be 10 revolutions per second, the E. M. F. set up between the brushes will be

$$\frac{10 \times 1{,}000 \times 1{,}000{,}000}{100{,}000{,}000} = 100 \quad \text{volts,}$$

and, since there are 25 commutator bars on each side of the diameter of commutation, there will be an average of four volts per coil of 20 turns. If the resistance of each coil be 0.01 ohm, the current which tends to be set up in a short-circuited coil having the average E. M. F. is

$$\frac{4}{0.01} = 400 \quad \text{amperes.}$$

201. It now remains to be explained how the existence of a powerful current in the short-circuited coil will produce violent sparking at the commutator. It is well known that the presence of a spark indicates a higher E. M. F. than the four volts, which we have assumed in this case is to be generated in the short-circuited coil. The increase in the voltage at the moment of sparking is due to what is called the *inductance* of the coil.

At the moment of short circuiting the coil by the bridging of the brushes across the two adjacent commutator segments, a powerful magnetic flux is set up in the coil, owing to its M. M. F. This flux is so directed through the coil as to set up in it an E. M. F. which opposes the development of the current. On the cessation of the current, owing to the breaking of the coil's circuit at the commutator, the coil is rapidly emptied of flux, and a powerful E. M. F. is set up in the same direction as the current, sufficiently powerful to produce sparking between the brush and the edge of the segment it is leaving. The E. M. F. so generated during the filling or emptying of the loop by its own flux is called the *E. M. F. of self-induction.*

202. Fig. 151 diagrammatically represents the flux produced in the short-circuited coils C', C, by the M. M. F. of the current produced during the short circuit. This flux passes principally through the air-gap and neighboring pole face, a small portion passing through the air in the interior of the armature between the core and the shaft. The greater the flux produced by the coil, the greater will be the E. M. F. developed in the coil, when the flux is suddenly withdrawn. The capability of a conducting loop or turn for producing E. M. F. by self-induction is called its inductance, and may be measured by the linkage of flux with the turn per ampere of the current it carries, that is, by the amount of flux passing through it.

203. We have thus far considered the coils C, C', as being composed of a single turn. If, however, each of these coils is composed of two turns, and the same current strength passes through each of these turns, then the M. M. F. of the coil will be doubled, and, if the iron in the armature core and pole face, is far from being saturated, the amount of flux passing through the two turns will be twice as great as that which previously passed through one. When this flux is introduced or removed it will generate E. M. F. in both turns, and, consequently, will induce twice as much E. M. F. in the two turns together as in a single turn. The inductance of the coil, or its capacity for developing E. M. F. by self-induction, is thus four times as great with two turns as with one, because there is

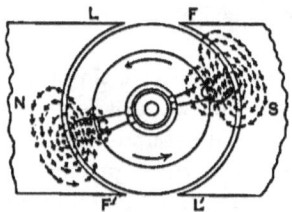

FIG. 151.—DIAGRAMMATIC REPRESENTATION OF FLUX IN MAGNETIC CIRCUIT OF SHORT-CIRCUITED COIL.

double the amount of flux, and a double the number of turns receiving that flux.

204. It is evident, therefore, that the inductance of a coil increases rapidly with the number of its turns, and although not quite proportionally to the square of the number, since, with a large number of turns, although the M. M. F. is increased in proportion to the number, yet the amount of flux passing through each of the turns, owing to leakage, is not the same. The E. M. F. of self-induction generated in each coil depends:

(1.) Upon the E. M. F. induced in the coil by the revolution of the armature.

(2.) Upon the resistance of the coil, or its capability for allowing a large current to flow through it.

(3.) Upon the inductance of the coil, or its capability for

permitting that current to induce a powerful E. M. F. when the circuit is made or broken.

The E. M. F. induced on making the circuit at the commutator is advantageous, since it checks the development of the current; the E. M. F. induced on breaking the circuit is harmful, since it enables a spark to follow the brush.

If, therefore, no sparking is to occur in a dynamo-electric machine at no load, the brushes must rest on segments, connected with coils in which no E. M. F. is being generated.

205. If the external circuit of the armature be closed through a resistance, so that current flows through the armature coils and brushes into the external circuit, the preceding conditions become considerably modified.

Fig. 152 represents the condition of affairs in which a current

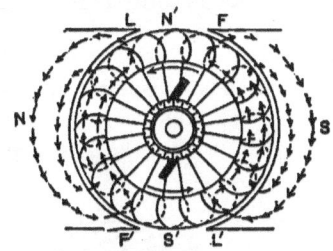

FIG. 152.—DIAGRAMMATIC REPRESENTATION OF MAGNETIC CIRCUIT OF ARMATURE.

is flowing through the armature coils, and the brushes are resting on the commutator, with the diameter of commutation at the neutral points, or in a plane at right angles to the polar axis.

In this figure the direction of the armature rotation is the same as shown in previous figures; namely, counter-clockwise. Here the flux produced by the M. M. F. of the armature coils takes place in the circuits digrammatically indicated by the curved arrows. The magnetization, therefore, produced by the current circulating in the armature turns, is a *cross magnetization*, or a magnetization at right angles to the magnetization set up by the field flux. The field flux through the poles and armature is diagrammatically indicated in Fig. 153, where the north pole is assumed to be situated at the left-hand side of the figure, and the average direction of the field flux is at right angles to the average direction of the armature flux. An inspection of Figs. 152 and 153 will show that at the *leading edges* of the pole-piece, L, L', that is, at those edges of the pole-piece which the armature is approaching, the direction of the flux produced by the armature is opposite to that of the

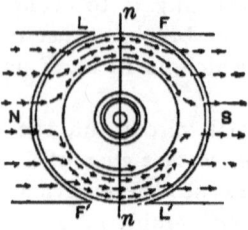

FIG. 153.—DIAGRAMMATIC REPRESENTATION OF FIELD FLUX PASSING THROUGH ARMATURE.

flux produced by the field, and that, consequently, the effect of superposing these fluxes is to weaken the flux at the leading edge as is shown in Fig. 154. On the contrary, at the *following edges* F' and F, of the pole-pieces, the direction of the armature

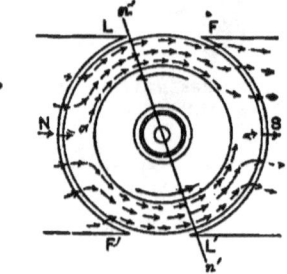

FIG. 154.—EFFECT OF SUPERPOSING ARMATURE FLUX ON FIELD FLUX.

flux coincides with the direction of the field flux, and the superposition of these two fluxes will have the effect of intensifying the flux at the following edges. Consequently, the *neutral line* in the armature, or the line symmetrically disposed as regards the entering and leaving flux, will no longer occupy the position N, N, at right angles to the polar axis, but will occupy a position n n'; therefore, in order to set the brushes so that they may rest upon commutator segments connected with coils having no E. M. F. generated in them, it is necessary to bring the diameter of commutation into coincidence with the neutral line, or to give the brushes a *lead; i. e.*, a forward motion, or in the direction in which the armature is rotating.

206. This, however, will not in itself, as a rule, prevent sparking, for the reason that induced E. M. Fs. are produced in the coil under commutation at load, even although the coil being commuted has no resultant E. M. F. set up by rotation. This induced E. M. F. is due to the inductance of the coil and

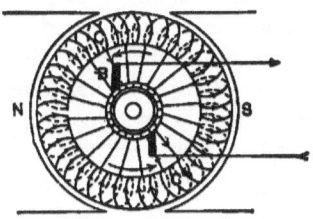

FIG. 155.—REVERSAL OF CURRENT IN ARMATURE COILS DURING COMMUTATION.

to the load current which it carries. An inspection of Fig. 155 will show that as the coil C, approaches the brush B, the current in the coil, as shown by the arrows, is directed upward on the side facing the observer; while on leaving the brush, after having undergone commutation, the current in the coil will be flowing in the opposite direction or downward. The sudden reversal of the current in the coil under commutation produces a sudden reversal of the magnetic flux linked with the local magnetic circuit of that coil, and this sudden change in the magnetic flux through the coil induces in it a powerful E. M. F., causing a spark to follow the brush.

In order that no spark shall be produced from this cause, it is necessary that before the brush leaves the segment the current in the coil shall have become reversed, and will therefore be flowing in the same direction as that which will pass through it during its passage before the pole face N. In order to effect this it is necessary to bring the coil that is being commutated into a field of sufficient intensity to induce in it, while short circuited, a current strength equal and opposite to that which passes when it first becomes short circuited by the brush. It is not, therefore, usually possible to keep the brushes on the neutral line as shown in Fig. 154, at $n\,n'$, but their lead must be increased, until the coil under commutation is in a sufficiently powerful field beneath the pole face to produce, or nearly produce, this reversal of current. The amount of lead necessary to give to the brushes in order to effect this will depend upon the inductance of the coils, and also on the strength of the current in the armature.

207. The lead of the brushes, besides tending to reduce sparking at the commutator, tends to diminish the E. M. F. generated by the armature, for two distinct reasons : First, because it connects in series armature windings in which the E. M. Fs. are in opposition, as will be seen from an examination of Fig. 156; and second, because the M. M. F. of the armature coils over which the lead has extended exerts a C. M. M. F. in the main magnetic circuit of the field coils, thereby tending to reduce the useful flux passing through the armature. This effect is called the *back-magnetization* of the armature. Cross-magnetization, therefore, exists in every armature as soon as it generates a current, but back-magnetization only exists when a current is generated in the armature, and the diameter of commutation is shifted from the neutral points.

208. The conditions which favor marked sparking at the commutator of a generator are, therefore, as follows; namely,

(1.) A powerful current in the armature; *i. e.*, the sparking increases with the load.

(2.) A large number of turns in each coil connected to the commutator; *i. e.*, the sparking increases with the inductance.

(3.) A great distortion of the neutral line through the armature, or a powerful armature reaction.

(4.) A high speed of rotation of the armature, since the current in the coil has less time in which to be reversed during the period of short circuiting.

(5.) A nearly closed magnetic circuit for each coil; *i. e.*, a small reluctance in the magnetic circuit of each coil, whereby the inductance of the coil is increased.

The conditions which favor quiet commutation, or the absence of sparking, are as follows; namely,

(1.) A small number of turns in each commuted coil, or a large number of commutator bars.

(2.) Decrease of current strength through the armature.

(3.) A lead of the brushes.

(4.) A powerful field, or a high magnetic intensity in the entrefer, due to the M. M. F. of the field magnets.

(5.) A large reluctance in the magnetic circuit of each coil.

209. An inspection of Figs. 152–154 will render it evident that the effect of superposition of the armature M. M. F. upon the M. M. F. of the field magnets, is to weaken the intensity of the field flux at the leading edges of the pole-pieces, and to strengthen the intensity at the following edges of the pole-pieces. At the same time, it is necessary to advance the brushes; $i.\,e.$, the diameter of commutation, so as to bring the commuted coils under the leading edges of the pole-pieces, in order that they may receive a sufficiently powerful intensity of field flux to enable the armature current to be reversed in the coil under the brushes, and sparkless commutation thus to be effected. If, however, the number of ampere-turns on the armature; $i.\,e.$, its M. M. F. at a given load, be sufficiently great, the field flux at the leading edges of the poles will be so far weakened, that the intensity left there will be insufficient to effect sparkless commutation, no matter how great the lead may be. In other words, the flux from the armature will overpower the field flux, in any position of the brushes. This will take place when the M. M. F. due to half the turns of active conductor on the armature, covered by the pole face, is equal to the drop of magnetic potential in the field flux through the entrefer.

210. The magnetic intensity under the edge of the leading pole-piece will be zero, when the magnetic difference of potential between this polar edge and the armature core, immediately beneath, is zero. The magnetic difference of potential across the gap at this point due to the field flux alone, will be the magnetic drop in the entrefer, or $\mathcal{B}d$, where $\mathcal{B}$, is the field intensity in the gap with no current in the armature, and d, the length of the entrefer in cms. The total M. M. F. of the armature, along the arc of one pole, will be $\frac{4\pi}{10} wp$, where wp is the number of turns covered by the pole, and this will be the total difference of potential in the magnetic circuit of the armature. Assuming that the armature is not operated near the intensity of magnetic saturation, almost the entire reluctance in the armature circuit will be in the entrefer. Fig. 156 represents diagrammatically the magnetic circuit of a Gramme-ring armature. The reluctance between bc and cd, in the field pole, also between ef and fa, in the armature, will be comparatively small, so that the total magnetic difference of potential developed by the armature will be expended in the two air-gaps ab and de, half the M. M. F. of the turns beneath the pole face being expended in each air-gap. Strictly speak-

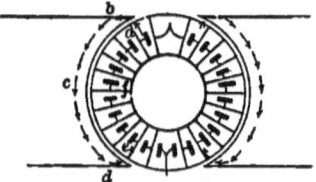

FIG. 156.—MAGNETIC CIRCUITS OF GRAMME-RING ARMATURE DUE TO ITS OWN M. M. F.

ing, the magnetic flux produced by the armature will not be confined to the paths indicated by the dotted arrows, but will pass across the air-gap at all points not situated on the neutral line cf. The above principles may be relied upon, however, to a first approximation.

211. In order, therefore, that a smooth-core armature should be capable of sparkless commutation, the M. M. F. of the turns on its surface, covered by each pole, should be somewhat less than the drop of magnetic potential in each air-gap, so as to leave a residual flux from the field in which to reverse the armature current in the coil under commutation. For example, if each air-gap or entrefer has a length of 2 cms., and the intensity in the air is 3,000 gausses, the drop of potential in the air will be 6,000 gilberts. If the number of

Gramme-ring armature turns, covered by one pole-piece, is 200, then a current of 80 amperes from the armature will represent 40 amperes on each side, and the M. M. F., produced by this current will be $\frac{4\pi}{10} \times 40 \times 200 = 10{,}056$ gilberts, and half of this amount, or 5,028, being less than the drop of field flux in the gap, should leave a margin for sparkless commutation.

212. Although the preceding rule enables the limit of current for sparkless commutation, on a smooth-core armature, to be predicted under the conditions described, yet it by no means follows that sparkless commutation must necessarily be obtained if the M. M. F. of the armature lies within this limit. If, for example, the number of commutator segments is very small, the inductance of each segment may be considerable, and a powerful flux intensity may be required to reverse the current under the brush in the presence of such inductance. No exact rules have yet been formulated for the determination of the inductance in a coil with which a given current strength, speed of rotation, and field intensity, shall render sparkless commutation possible.

213. The methods in general use for the suppression of sparking may be classified as follows:

(1.) Those which aim at the reduction of inductance in the commuted coils.

(2.) Those which aim at the reduction of the current strength passing through the coil during its short circuit by the brush, and, therefore, at the reduction of the current strength which must be reversed before the short circuit is over.

(3.) Those which aim at the reduction of the armature reaction, so as to reduce its influence in weakening the field intensity in which the coil is commuted.

214. There are two methods for reducing the inductance of the armature coils.

The first is to employ a great number of commutator segments, thus decreasing the number of turns in each coil under commutation. It is evident that an indefinitely great number of commutator segments would absolutely prevent sparking. A great number of commutator segments is, however, both troublesome and expensive, so that in practice a reasonable maximum cannot be exceeded.

The second method for lessening the inductance of the armature coils differs from the preceding only in the method of connection. It consists in providing two separate windings or sets of coils ; or, as it is sometimes called, in *double-winding* the armature. The two separate windings are insulated from each other, but are connected to the commutator at alternate segments, so that the brush rests coincidently upon segments that are connected with each winding. Each winding therefore, furnishes half the current strength, and the effect of the inductance in each coil is reduced.

215. When the brushes are not so shifted as to bring the diameter of commutation into coincidence with, or even in advance of, the neutral point, the coil under commutation will be situated in a magnetic flux in the wrong direction; *i. e.*, a magnetic flux which tends to increase and not to reverse the current strength in the coil, so that when the coil is short circuited by the brush, the current strength becomes increased in the wrong direction. When, for any reason, it is impossible to alter the lead of the brushes during variations of load, as, for example, when the generator has to run without attendance, the sparking, which may be produced at the brushes owing to the resultant flux in which the commuted coils lie, may be greater than that due to the mere reversal of armature current in the coil under the influence of its inductance. In such cases, considerable improvement is effected by the insertion of a resistance between the coils and the commutator segments with which they are connected. Thus in Fig. 157, the connecting wires a_2 and c_1, are sometimes made of German silver. It is evident, under these circumstances, that the coil undergoing commutation will not only have its own resistance, but also the resistance of the German silver wires in the local circuit through the brush, and the current which can be set up in this circuit by the E. M. F. induced in the coil, owing to its motion through the distorted field, is prevented from assuming considerable strength. The value of the German silver resistances, although great by comparison with the resistance of a single coil, is small when compared with the resistance of the entire armature, and, consequently, does not greatly add to the armature's effective resistance. It is clear that this method does not obviate the sparking due to the inductance of the armature coils, but tends rather to obviate that due to the establishment of unduly powerful currents in the short circuited-coil in the wrong direction, and which current has suddenly to be reversed when the short circuit is broken. The method is, therefore, often employed with armatures for which the brushes cannot be shifted.

216. The most generally adopted plan for reducing sparking is to employ a comparatively high resistance in the brush

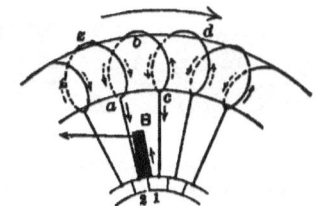

FIG. 157.—CURRENT FLOW IN ARMATURE COIL UNDER COMMUTATION.

itself. An examination of Fig. 157, will show that if the resistance in the tip of the brush B, can be made sufficiently great, the current which enters the commutator from the wires will be so far reduced, before contact with the brush tip ceases, that when the rupture does take place, practically all the current from the armature will be passing through the coil in the right direction; *i. e.*, in the same direction as it will have when the brush has passed to the next coil, and, consequently, the current strength which has suddenly to be reversed when the brush leaves the bar is very small.

217. Thus in Fig. 157, suppose the armature is rotating in the direction of the large curved arrow, and that the commutator segment 1, is about to move from beneath the brush B. The coil 2 *a b c* 1, is about to change position, from the left-hand to the right-hand side of the armature, and the current in the coil is about to change in direction, as indicated by the small curved arrows, from the direction *a b c*, to the direction *c b a*. The current leaving the armature having recently been flowing to the brush *B*, from section 1, and the wire *c* 1, is now flowing from sections 2 and 1, and from wires *a* 2 and *c* 1. If the resistance in the tip of the brush is considerable, relatively to that in the whole breadth of the brush, the current through *c* 1 *B*, will be relatively reduced and that through *a* 2 *B*, relatively increased. This will require, however, that the current from the right-hand side of the armature shall be forced through the coil *b*, in the direction *c b a*, and the more nearly this can be accomplished, before contact is broken between 1 and *B*, the less is the opportunity that is offered for the inductance of the coil *a b c*, to produce a spark at rupture. With this pur-

FIG. 158.—DYNAMO BRUSH OF STRIPS OF INTERLEAVED COPPER AND HIGH RESISTIVITY METAL.

pose in view, brushes are made up of strips of German silver, interleaved with copper or woven gauze; or they may be made of carbon with a specially high resistivity. Fig. 158 represents a form of brush in which strips of copper are interleaved between strips of high resistivity metal. By this means the brush, as a whole, possesses the requisite conductance for the current it has to carry, but the tip has sufficient resistance to assist in the reversal of the current in the coil under commutation. Fig. 159 represents a block of carbon employed in a suitable holder or frame as a dynamo brush. In order to increase the conductance of the brush as a whole, it is usually thinly copper-plated as shown. Carbon brushes are largely employed for 120-volt dynamos where the current strength produced is not great, and almost exclusively employed with 500-volt dynamos. The use of such brushes tends to reverse the current in the armature, during the period of short circuiting, and also aids in checking any undue current in the wrong direction, caused by distortion of the field flux, owing to armature reaction.

Artifically compressed graphite is sometimes used for dynamo brushes. Besides the advantage of high resistivity, it lubricates the commutator surface.

218. Referring now to the third method for suppressing sparking at the commutator, a variety of plans have been attempted at different times for bringing about a reversal of the current in a commuted coil, during the period of short circuiting, by the action of a specially directed magnetic flux upon this coil, as, for example, by winding a special magnet

FIG. 159.—CARBON DYNAMO BRUSH.

placed with its pole immediately over the short-circuited coil, in such a manner that the flux from this magnet, penetrating the moving coil under commutation, may induce in it an E. M. F. sufficiently powerful, to set up in the short circuit, a current strength equal to that which the coil must sustain after commutation is over, or, in other words, to produce automatically the same effect which the lead of the brushes would be capable of effecting under the most favorable conditions. When, however, the current through the armature and its M. M. F. are powerful, the M. M. F. needed on such controlling magnets may require to be very considerable, and, for this reason, the plan, in this form, has never come into general use.

219. In the same direction a method has recently been proposed for obtaining sparkless commutation by introducing into the magnetic circuit of the machine, a M. M. F. equal in amount, but opposite in direction, to that of the armature. This has the effect of practically preventing all armature reaction and distortion of the field flux. It is carried out by winding around the armature and through the field poles, as shown in Fig. 160, a number of turns, between A and B, equal to that of the armature winding, and in series with the armature, so that the ampere-turns in the *balancing coil A B*, are equal and opposed to the ampere-turns on the armature. The two M. M. Fs. thus counterbalance and neutralize each other, leaving the field flux practically unchanged at all loads of the

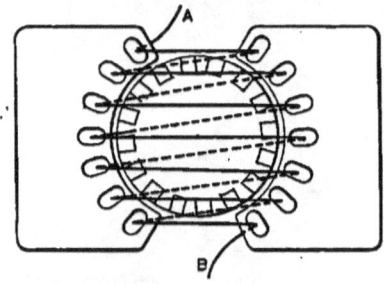

FIG. 160.—DEVICE FOR PREVENTING ARMATURE REACTION.

machine. By this means all sparking due to distortion of the field is prevented, and only the sparking due to the inductance of the commuted coil, and the current reversal in the same, is left. In order to check this, an additional winding or magnet over the commuted coil is introduced for the purpose of reversing the E. M. F. in this coil as above described, a process which is more easy of application when no armature reaction exists than when armature reaction is unchecked. A quadripolar machine, wound in this manner with a quadruple set of balancing coils, is shown in Fig. 161.

220. While it is claimed for this method that it entirely overcomes armature reaction, yet it possesses the disadvantage that it requires the use of what is practically an extra armature winding upon a part of the machine which does not revolve, thus introducing an additional cost and complexity. It, therefore, remains to be determined how far the advantage of sparkless operation is offset by extra resistance, weight, material, and cost.

Another method, which has been tried in England for the purpose of suppressing sparking, adds extra coils on the armature, one between each commutator segment and its

FIG. 161.—QUADRIPOLAR GENERATOR WITH BALANCING COILS.

armature connection. These coils are arranged in such a manner that the E. M. F. induced in them by their revolution through the field shall reverse the direction of the current in the coil under commutation. Fig. 162 represents diagrammatically the method of winding, and Fig. 163 the action of the various E. M. Fs. In Fig. 162 the inner ring with the additional coils actually forms part of the armature core and receives the flux from the field although indicated in the figure as a separate ring for clearness of description. Fig. 163 shows a coil being short circuited by the brush, and the direction of the current in this coil is being reversed by the action of its auxiliary coil which is still under the trailing pole edge, so that when the bar B, leaves the brush, no serious spark shall follow.

221. In the dynamo-electric machine represented in Fig. 6, and which has but three commutator segments, the spark is

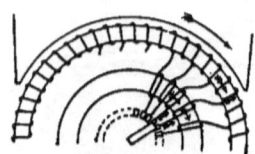

FIG. 162.—DIAGRAM OF CONNECTIONS OF EXTRA ARMATURE COILS FOR CHECKING SPARKING.

prevented from forming by an air blast directed against the commutator in such a manner as to extinguish the incipient spark at the breaking of the short circuit. This air blast is

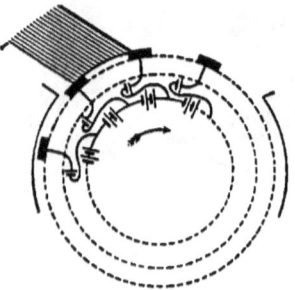

FIG. 163.—DIAGRAM INDICATING ACTION OF DEVICE ILLUSTRATED IN FIG. 156.

supplied by a small centrifugal pump rotating with the armature.

222. The number of bars in the commutator of a generator depends principally upon the sparking limit. If there were no danger of excessive sparking, the number of commutator bars in any machine would be very small, except when marked freedom from pulsation is required in the current strength.

The number of bars will, therefore, depend upon the pressure and current strength, the armature reaction, and the field flux intensity. An unduly small number of bars leads to excessive sparking, and, in the case of high pressure machines, the sparks may flash completely around the commutator, producing what is practically a short circuit. Small machines have been built, however, giving 10,000 volts with only 32 commutator segments.

223. Thus far we have mainly considered smooth-core armatures. The great majority of dynamos, in construction at the present time are, however, toothed-core armatures. In the first production of toothed-core machines, the sparking which they exhibited was more troublesome and violent than in smooth-core armatures of equal size, and apparently for the reason that the inductance of each armature coil was increased, owing to its being surrounded, or nearly surrounded, by iron, instead of having an iron base only, as in the smooth-core type. This difficulty has since been overcome by careful designing, and toothed-core armatures are now constructed which give less trouble from sparking than smooth-core armatures of equal size and output. This is accomplished by giving such a cross-section to the teeth in the armature that, at no load, the iron in the teeth is nearly saturated, and has, therefore, a high reluctivity. The presence of armature reaction tends to increase the magnetic intensity in the teeth beneath the trailing pole edges, and to diminish it in the teeth beneath the leading pole edges, as already observed. This tendency is opposed by the increasing reluctivity of the saturated teeth at the trailing pole edges, and, consequently, the teeth tend to restore an equal distribution of magnetic flux over the surface of the armature; or, in other words, tend to check the effect of armature reaction. At the same time, the high reluctivity of the teeth tends to diminish the inductance of each coil undergoing commutation, so that, by careful adjustment, the existence of the teeth is not merely a mechanical advantage but also a considerable electrical advantage.

In practice, the output of a generator is not really limited by excessive sparking. As usually designed, the temperature elevation of the armature, even when thoroughly ventilated, fixes the limit to the output before the sparking becomes troublesome. And, in fact, many generators are in use to-day which never require to have any lead given to their brushes, and need only occasional attention to their commutators.

224. In the preceding discussion, we have considered armature reaction from the standpoint of the Gramme-ring armature only, but the same principles are equally applicable to disc or drum armatures.

CHAPTER XIX.

HEATING OF DYNAMOS.

225. The activity expended in any generator invariably takes the form of heat. These expenditures are:

(1.) $I^2 R$ activity in the field magnets.
(2.) $I^2 R$ activity in the armature winding.
(3.) $I^2 R$ activity in eddy currents, in armature and field.
(4.) Hysteretic losses in armature core and field poles.
(5.) Friction in bearings and brushes.
(6.) Friction in air.

226. The number of watts expended in the field magnets is equal to the product of the pressure in volts at the field terminals, multiplied by the current in amperes passing through the field. This activity, although steadily expended in the form of heat, is necessary in order to produce the M. M. F. of the field-coils. In a certain sense, therefore, it may be said that the $I^2 R$ activity in the field windings is expended in order to magnetize the field, and the $I^2 R$ activity in the armature winding is expended in order to magnetize the armature. In series-wound generators, where the armature sends its entire current through the field magnets, this expenditure varies with the load. Thus, in a 10-KW series-wound generator, designed to supply a maximum current of 200 amperes at 50 volts pressure, if the resistance of the field magnet coils, when warm, be 0.01 ohm, the pressure at the terminals of the magnets will be 200 × 0.01 = 2 volts, and the activity, 2 × 200 = 400 watts. On light load, however, of say 20 amperes, the pressure will be 0.01 × 20 = 0.2 volt, and the activity 0.2 × 20 = 4 watts, so that, in the first case, the amount of heat generated in the field winding is 100 times greater than in the second case, and the temperature, which the field winding would attain in the first case, would be much higher than in the second.

227. In a shunt-wound generator, the activity in the field circuit is nearly constant. For example, a 10-KW generator, intended to supply 111 volts at its terminals, at full load, with a current strength of 90 amperes in the main circuit, might supply a current of 2.5 amperes through its field magnets. Consequently, the activity in the field-magnet circuit would be $111 \times 2.5 = 277.5$ watts. At light load, the current strength through the field magnets would have to be reduced to say 2.0 amperes, in order to keep the terminal pressure at 111 volts, and the activity in the field would be reduced to $111 \times 2.0 = 222$ watts, so that the temperature attained by the winding on the field magnets would not be much greater at full load than at no load.

228. It is evident that the $I^2 R$ activity in the armature always varies with the load; *i. e.*, with the current strength I. At no load, this loss must be very small, the current strength being limited to that required for the excitation of the field magnets. The temperature elevation of the armature, due to the armature winding, consequently, increases rapidly with the load.

229. The activity expended as $I^2 R$, in eddy currents in the field poles, or in the armature, is nearly uniform at all loads, especially in shunt-wound machines, in which the intensity of magnetic flux is nearly constant, and if this intensity were absolutely uniform; *i. e.*, if there were no drop in the armature, requiring a greater M. M. F. and exciting current, and if there were no armature reaction, the eddy current loss would be constant at all loads.

230. The activity expended in hysteresis in the armature and field poles, would, similarly, be constant at all loads if the magnetic intensity were constant. As the magnetic intensity is increased by an increase in the M. M. F. of field and armature at full load, the hysteretic loss increases, approximately following the 1.6th power of the local magnetic intensity at any point. The heat due to hysteretic loss is developed principally in the armature.

231. The friction in bearings and brushes produces heat at those parts. The amount of heat liberated, due to pure friction, is comparatively small when the lubrication of the bearings is properly attended to. In large generators, the heat produced by the friction of the brushes on the commutator is very small compared with the heat developed by the sparking, and the powerful currents set up in the short circuited coils undergoing commutation.

The frictional forces opposing the rotation of an armature in which there is no appreciable magnetic flux, are due to gravitation; *i. e.*, to the weight of the revolving parts. When, however, the field magnets are excited, and magnetic flux passes through the armature, the frictional forces are due to gravitation and magnetic attraction combined. If the armature is situated symmetrically with respect to an external system of field magnets, if for example, the Gramme-ring armature of Fig. 129 be revolved concentrically with the polar bore, the system of magnetic forces all round the machine will balance, and the friction of the machine will not be increased by the influence of the magnetic flux. If, however, the armature were nearer the lower poles, so that the entrefer was shorter beneath the armature than above it, there would be a tendency, as we have seen, to produce a greater magnetic intensity in the lower magnetic circuits than in the upper ones, with a corresponding resultant magnetic pull upon the armature, vertically downward. The armature would consequently revolve in its bearings as though its weight were increased, and with an increase in friction and frictional expenditure of energy. On the other hand if the armature were centred too high, so as to develop greater magnetic fluxes in the upper than in the lower magnetic circuits, the effective weight of the armature in its journals would be reduced, and the frictional waste of energy in them diminished. This principle has been employed in the design of some bipolar machines, in which the resultant magnetic attraction upon the surface of the armature is upwards, or in opposition to the attraction of gravitation.

232. The friction due to the churning of the air is comparatively small in drum armatures, but often constitutes an appreciable loss in alternators, when a Gramme-ring armature of large diameter and rough exterior is revolved at a high speed. In this friction the heat is principally developed in the surrounding air and not in the mass of the machine. The air churning, on the contrary, assists in cooling the machine.

233. The magnetic stresses exerted in large electro-dynamic machines are often of considerable amount. Referring for example to the machine outlined in Fig. 129, the polar areas are 1,400 sq. cms., and the useful magnetic flux passing perpendicularly into the armature, 3.534 megawebers. The mean intensity in the entrefer is therefore $\frac{3,534,000}{1,400} = 2,524$ gausses. The attractive force per square centimetre (Par. 72) is $\frac{B^2}{8\pi} = \frac{2,524 \times 2,524}{25.133} = 253,400$ dynes $= 258.4$ grammes. The total stress exerted will be $1,400 \times 258.4 = 361.700$ grammes $= 797.4$ lbs. weight, at each pole.

234. In drum or Gramme-ring armatures with radial field magnets, the magnetic flux through the armature, can only alter, within certain limits, the vertical forces acting upon the armature due to gravitation. In machines with parallel field magnets, as for example, in the dynamo of Fig. 8, the magnetic stresses exerted upon the armature are side thrusts, or horizontal stresses parallel to the axis of the shaft. If the entrefer on each side of the armature has the same length, the two resultant magnetic forces exerted upon the armature will be equal, but if the armature is nearer one set of poles than the other, so as to produce a shorter entrefer on one side than on the other, there will be a tendency to produce a resultant side thrust toward the side of shorter entrefer. It is important, therefore, that generators of this type should have their armatures nearly midway between the polar faces.

235. The expenditure of energy as heat in a generator is objectionable, first, because it represents loss of power, and, consequently, reduced efficiency. Ten per cent. of loss in the generator due to all these causes combined, means approximately 10 per cent. more coal, 10 per cent. more water, and engines and boilers larger by 10 per cent. to supply a given electric activity, than would be necessary if it were possible to avoid these losses entirely; and second, because the heat developed may raise the temperature of the generator to an objectionably high degree and ultimately limit its output.

236. There are four limitations to the output of a continuous-current generator; viz.,

(1.) Insufficient mechanical strength to withstand the mechanical forces brought into play.

(2.) Insufficient efficiency, or insufficient electric pressure at the brushes, under load.

(3.) Excessive sparking.

(4.) Excessive heating.

The first two cases of limitation can always, by proper design, be obviated in all but the smallest generators. It is the third and fourth considerations which limit the output in all practical cases. In modern machinery it is the heating which first limits the output.

237. The limiting temperature of the generator armature is dependent upon a variety of considerations. In the first place, the hotter the armature winding becomes, the greater its resistance; for, if r, be the resistance of the armature, in ohms, at 0° C, its resistance R, at any temperature $t°$ C., will be approximately, $R = r (1 + 0.004\, t)$. In other words, the resistance will rise by 0.4 per cent. per degree centigrade of temperature elevation above zero. The result is, that at high temperatures, the wasteful activity, as I^2R, in the armature, increases, increasing thereby both the loss in the machine and the tendency to temperature elevation.

238. The temperature of the armature must not exceed that at which any of the materials employed in its construction would be deleteriously affected; *i. e.*, either softened or decomposed. In many generator armatures, cotton is the insulator employed, four thicknesses of cotton (representing each about $\frac{1}{130}$th of an inch, separates adjacent wires, except at specially protected places, where mica and oil paper are employed.

Cotton undergoes slow *thermolysis*, or decomposition by heat, at a temperature, approximately, that of the boiling point of water, or 100° C. Consequently, it is unsafe, in practice, to maintain cotton covered armatures, even though shellac-varnished, at a higher temperature than 100° C. If the temperature of the room, in which a generator is operated, never exceeded 30° C., it would require an elevation of 70° C. in the armature to reach a dangerously high temperature. As, however, some engine rooms attain, in summer, a higher temperature than 30° C., and since a margin has to be left for accidental overloads, 50° C. is the temperature elevation that the armature should not exceed at full load, and modern practice is reducing this to 40° C.; so that the temperature of the armature, as observed after several hours of full load, is usually specified not to exceed 40° C. of temperature elevation above the surrounding air.

United States Navy specifications usually require that the elevation of temperature shall not exceed 50° F. $= 27.8°$ C., at any part of the machine. Other things being equal, these specifications can only be met by increasing the size of machine for a given output. In other words, with machines of the same grade, a reduction of the limiting temperature at full load means a reduction of the load which the machine can carry.

239. Many large generators, however, do not use any insulation for their armature conductors, except mica, and such generators can safely carry a much higher temperature elevation without danger.

Here the dangerous temperature, so far as mechanical injury of the armature is concerned, would be that at which solder would melt. Electrically, however, the increase in the resistance in the armature would, probably, constitute a limitation long before this temperature was reached, and if, in fact, the armature winding were to attain this temperature, the field coils, and even the bearings of the machine, might be dangerously overheated.

240. The activity in the field coils, which will elevate their external temperature a given number of degrees centigrade, depends upon their shape, size and arrangement, whether their surfaces are freely exposed to the air, or are partly sheltered from it. Usually, however, the surfaces of the field coils must afford 16 square centimetres, or about 2.5 square inches per watt of activity developed in them as $I^2 R$ heat. If the field winding consists of many layers of fine wire, the temperature of the deep seated layers will be greater than that of the superficial layer; but if, on the contrary, the layers be few, and the wire coarse, the difference of temperature in the winding will be inconsiderable. The elevation of temperature on the field magnets of a generator is usually not greater than 30° C. at full load.

241. In the case of the armature, the speed at which it revolves through the air greatly increases its capability for dissipating heat and reducing its temperature, so that a much greater surface thermal activity can be permitted in the armature than in the field coils. The usual allowance for eddy currents, load currents and hysteretic losses combined, is about $\frac{1}{5}$ th watt per square centimetre; *i. e.*, $1\frac{1}{3}$ watts per square inch of armature surface, including the surface on the sides of the armature, but excluding its internal core surface; or, about three times more activity per unit area than on the field magnets. In some specially ventilated armatures, in which the core discs are spaced and separated at intervals, to permit the circulation of air from the interior outward by centrifugal force, the dissipation of heat can be so far increased that two watts per square inch of armature surface have been rendered practicable. Much depends, however, upon the shape and size of the armature, as well as upon its peripheral speed, so that no exact rule can be laid down.

CHAPTER XX.

REGULATION OF DYNAMOS.

242. As has already been pointed out (Par. 16), all self-exciting continuous-current generators may be wound in one of three ways; namely,
 (1.) Series-wound.
 (2.) Shunt-wound.
 (3.) Compound-wound.

243. Fig. 164 represents diagrammatically the connections between the field and armature of a series-wound generator.

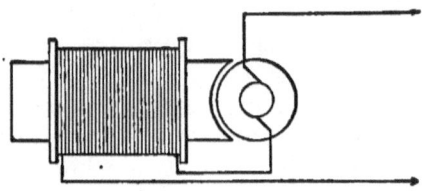

FIG. 164.—DIAGRAM OF SERIES WINDING.

It will be observed that the current in the main circuit passes through the field magnet windings. The M. M. F. of the field coils, therefore, increases directly with the current strength through the circuit. So long as the iron in the magnetic circuit of the machine is far from being saturated, the flux through the armature increases with the M. M. F., approximately, in direct proportion, and the E. M. F. of the armature, consequently, increases nearly in proportion to the current strength. As soon as the iron in the circuit approaches saturation, the flux increases more slowly, and finally, the E. M. F. of the armature is scarcely increased by any increase in the current strength through the circuit.

244. Fig. 165 represents diagrammatically the connections between the field and armature of a shunt-wound generator.

REGULATION OF DYNAMOS.

Here the field magnets are wound with fine wire and the windings are connected in parallel with the external circuit, instead of being connected in series with it. Consequently, if the pressure at the brushes be considered as uniform, the current strength passing through the magnet coils must, by Ohm's law, be uniform, independent of the current strength in the main circuit. Thus, if the pressure at the brushes be assumed constant, at, say 100 volts, and the resistance of the magnet coils be 50 ohms, then the current strength through the magnet coils will be two amperes, independently of the strength of current supplied to the main circuit.

245. Practically, however, owing to the drop of pressure in the armature as the load increases, and also on account of the

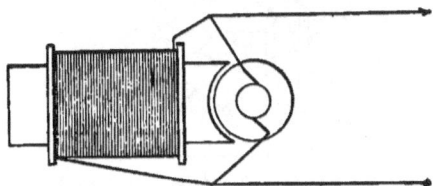

FIG. 165.—DIAGRAM OF SHUNT WINDING.

shifting of the brushes that may be necessary with the increase of load, the pressure at the brushes diminishes, and the current strength through the field magnets diminishes in the same proportion. The tendency in a shunt-wound machine is, therefore, to diminish its M. M. F., and its resulting E. M. F., as the load on the generator increases. In order to maintain a constant pressure at the brushes under all variations of load, it is necessary to adjust the strength of current passing through the field magnets, so that the M. M. F. at full load shall be slightly in excess of the M. M. F. at light load. This is usually accomplished by the insertion of a rheostat in the field magnet circuit, so that some or all of this resistance can be cut out by hand at full load, thereby increasing the current strength through the magnet coils.

246. If, for example, the full-load activity of the machine be 10 KW at 100 volts pressure, the full-load current strength will be 100 amperes. Assuming the resistance of the armature to be 0.05 ohm, the drop of pressure in the armature at full load will be $100 \times 0.05 = 5$ volts, and the additional drop of pressure, owing to the shifting of the brushes in order to avoid sparking, may be 2 volts more, making a total drop in pressure of 7 volts. The effect of this drop would be to reduce the current strength in the field magnet coils from 2 amperes to $\frac{93}{50} = 1.86$ amperes, thus reducing both the flux through the armature and the E. M. F., so that a balance between the E. M. F. and its excitation might be found at, say, 90 volts, if no means were adopted to regulate the current strength through the field coils. In other words, the

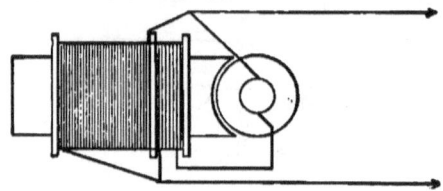

FIG. 166.—DIAGRAM OF COMPOUND WINDING.

pressure at the brushes would vary by 10 volts between light and full load.

247. Fig. 166 represents the connections between the field and armature of a compound-wound generator. Here the principal M. M. F. furnished by the magnet coils is that due to the shunt coil, composed of many turns of fine wire, an auxiliary series coil, of comparatively few turns of coarse wire, being also employed in the main circuit. As the load increases, the M. M. F. generated by the shunt winding tends to diminish as above described, but the M. M. F. due to the series coil increases. By suitably proportioning these two opposite influences, the M. M. F. may be automatically so controlled, that the pressure at the brushes shall remain constant, either at the brushes of the generator, or at the terminals of the motor or other translating device, which may be situated at a considerable distance from the generator. In order to effect this latter result, the M. M. F. of the series coil must compensate not only for the drop in the armature, but also for the drop in the conductors leading from the generator to the motor, so that these external conductors may be regarded, electrically, as forming an extension of the armature winding, and, in this sense, the generator delivers a constant pressure at its final terminals on the motor. Such a machine is said to be *overcompounded*.

248. Series-wound generators are almost invariably employed for series-arc lighting, since it would be very difficult to supply the required M. M. F. for their magnets by a shunt winding, considering that the pressure at the brushes varies between such wide limits; and, even if such shunt winding could be supplied, it would necessarily be formed of a very long and fine wire, and, consequently, would become troublesome and expensive. Series arc-lighting generators are sometimes constructed for as many as 200 lights, representing about 10,000 volts at the generator terminals at full load, and a shunt winding for such a pressure would be very expensive.

249. Shunt-wound generators are usually employed for supplying incandescent lighting from a central station, and their pressure is varied by hand regulation.

Compound-wound generators are usually employed for supplying motors from central stations, and also for incandescent lights and motors in isolated plants.

250. In the design and use of generators, it is important to know how the E. M. F. generated in the armature at a given speed varies with the current passing through the field magnets. We have seen that so long as the brushes remained unaltered in position, the E. M. F. in the armature, in C. G. S. units, is equal to the product of the number of turns on the armature, the number of useful webers passing through the armature from each pole, and the number of revolutions per second. Consequently, the E. M. F. of such an armature, running at a constant speed, depends directly upon the flux through its magnetic circuit or circuits. If we vary the current strength through the field magnets, and, consequently, the M. M. F., we can observe the pressure in volts, which the machine will deliver at its brushes at light load. A series of such observations, plotted in a curve, gives what is called the *characteristic curve* of the generator. In the case of a self-exciting, series-wound generator, it is only possible to

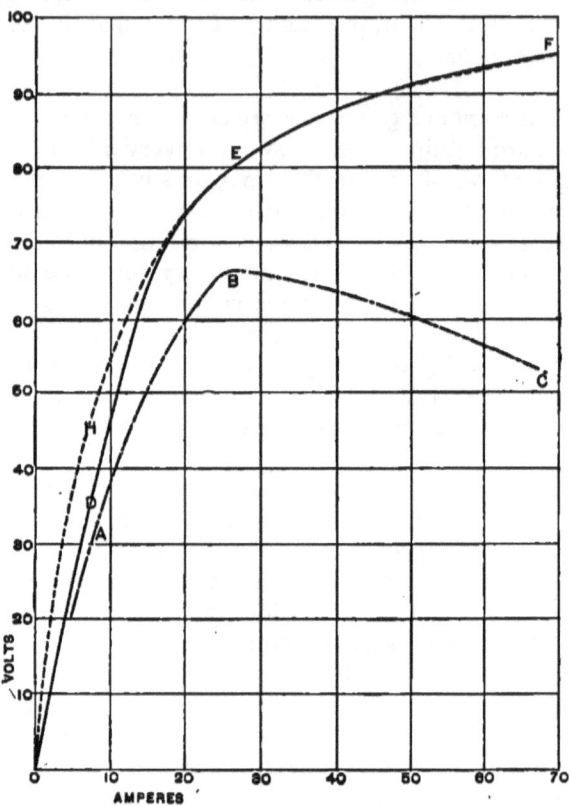

FIG. 167.—CURVE OF E. M. F. DEVELOPED IN THE ARMATURE OF A SERIES-WOUND DYNAMO, WITH REFERENCE TO CURRENT STRENGTH IN A CIRCUIT.

vary the M. M. F. by varying the load, and, consequently, by including, in the pressure at the brushes, the drop taking place in the armature. The curve obtained from a series-wound machine under such circumstances, is called an *external characteristic*, and the *internal characteristic* may be determined from it by correcting for the drop in the armature.

251. Fig. 167 represents the internal and external characteristics of a particular series-wound generator intended to supply a maximum of 70 amperes at 50 volts terminal pressure or 3,500 watts.

The pressure at terminals, when the load was varied so as to produce the required variations of current strength through the magnets, followed the broken line $A\cdot B\, C$, which is, therefore, the external characteristic of the machine. If we add to the ordinates of this line from point to point, the drop of pressure in the armature at the corresponding current strength, the full line o $D\, E\, F$, is obtained, which is, therefore, the internal characteristic of the generator or the curve of its E. M. F. in relation to the exciting current in its field coils.

The useful E. M. F. developed by the armature may be expressed by the formula,

$$E = \frac{I}{x + y I} \quad \text{volts.}$$

so that, if two observations are secured, the whole internal characteristic curve may be deduced to a very fair degree of accuracy. For example, in Fig. 167, the E. M. F. at 20 amperes $= 74$ volts, and at 70 amperes, 95 volts. From these observations we may take the two equations,

$$74 = \frac{20}{x + 20\, y} \text{ and } 95 = \frac{70}{x + 70\, y}.$$

From these two equations we obtain $x = 0.0836$ and $y = 0.00933$, so that the E. M. F. at any current strength through the field magnets is

$$E = \frac{I}{0.0836 + 0.00933\, I} \quad \text{volts.}$$

The dotted curve o $H\, E\, F$, which lies close to the full curve o $D\, E\, F$, represents the locus of this equation. It will be observed that the dotted line practically coincides with the full line representing the observations, except within the first 20 amperes of magnetizing current strength.

252. Fig. 168 represents the characteristic curve of a shunt-wound generator, of 200 KW capacity. Here the current strength through the field magnets was not observed, but the pressure acting on the field coils was noted. Assuming, as would probably be very nearly true, that the resistance of the field magnet coils remained constant throughout the observations, the exciting current strength would be proportional to the pressure acting on the coils. With 40 volts on the magnets, the E. M. F. at the brushes with the external circuit broken was 71 volts, and increased, as shown by the full line ABC, to

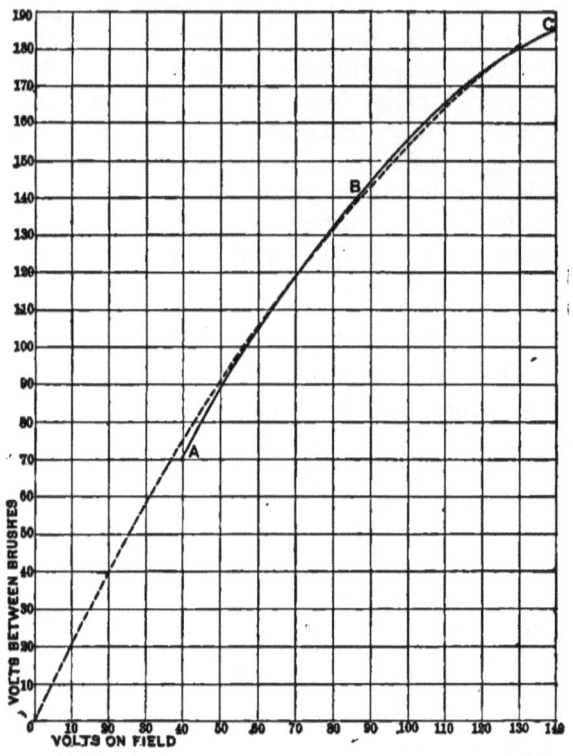

FIG. 168.—CHARACTERISTIC CURVE OF SHUNT-WOUND DYNAMO.

185 volts, with 140 volts on the magnets. Here also the E. M. F., E, may be expressed by the Frölich equation, $E = \dfrac{e}{x + ye}$, e being the pressure on the field magnets; taking the two observations, $120 = \dfrac{70}{x + 70\,y}$ and $174 = \dfrac{120}{x + 120\,y}$, we find $x = 0.43$ and $y = 0.0022$, from which the general equation becomes,
$$E = \dfrac{e}{0.43 + 0.0022\,e} \quad \text{volts.}$$

The locus of this equation is represented by the dotted line, which practically coincides with the full line $A\ B\ C$, of observation.

253. When, therefore, two reliable observations have been made of the E. M. F. generated by an armature, at observed exciting current strengths, or pressures, situated not too closely together, it is possible to construct the characteristic curve throughout to a degree of accuracy sufficient for all practical purposes.

The Frölich equation, by which this is possible, is a consequence of the fact that the reluctance of the air paths in the magnetic circuit of a generator is constant, while the reluctivity of the iron in the circuit is everywhere capable of being expressed by the formula $\nu = a + b\mathcal{H}$ (Par. 59); and, consequently, the total apparent reluctance of the armature takes the form $x + y\mathcal{F}$, and the useful flux passing through the armature $\Phi = \dfrac{\mathcal{F}}{x + y\mathcal{F}}$, $\mathcal{F}$, being the magnetomotive force in gilberts, but $\mathcal{F}$, may be expressed in ampere-turns, in amperes or in volts applied to the coils.

254. When the characteristic curves of a shunt machine have been obtained, it is a simple matter to determine what the series winding must be in order to properly compound it, either for the drop in the armature, or for the drop in any given portion of the external circuit as well. Thus, suppose it be required to determine the series winding for the machine whose characteristic curve is represented in Fig. 168. If the E. M. F. required at the terminals of the machine be 120 volts at all loads, and if the drop in the armature, due to its resistance at full load, as well as the resistance of its series coil, and to any shifting of the brushes that may be necessary, amounts in all to 10 volts, then the full-load current must supply the M. M. F. necessary to carry the E. M. F. from 120 to 130 volts, equivalent to raising the pressure by 8 volts from 70 to 78 volts on the shunt winding. The increase in current strength from the shunt winding represented by these eight volts multiplied by the number of turns in the shunt winding, gives the M. M. F. required, and the full-load current must pass through a sufficient number of turns to supply this M. M. F. in its series coil.

255. In all commercial circuits, electro-receptive devices require to be operated either at constant current or at constant pressure. The majority of such devices are designed for constant pressure; such, for example, are parallel or multiple-connected incandescent lamps and motors. Some devices, however, require to be operated by a constant current. Of these, the arc lamp is, perhaps, the most important. Series-

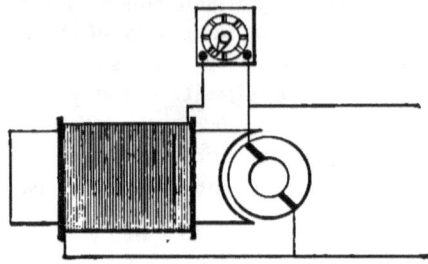

FIG. 169.—SHUNT FIELD AND RHEOSTAT.

connected incandescent lamps, and a few forms of motors, also belong to this class.

256. In order to maintain a constant pressure at the terminals of a motor with a varying load, it is necessary, in order to compensate for the drop of pressure in supply conductors, that the pressure at the generator terminals either be kept constant, or slightly raised as the load increases. With shunt-wound machines this regulation requires to be carried out by hand, a rheostat being inserted between the field and the armature, as shown in Fig. 169.

257. Various forms are given to rheostats for such purposes. They consist, however, essentially of coils of wire, usually iron wire, so arranged as to expose a sufficiently large surface to the surrounding air, as to enable them to keep within safe limits of temperature under all conditions of use. The resistance is divided into a number of separate coils and the terminals of these are connected to brass plates usually arranged

in circles, upon the external surface of a plate of slate, wood
or other non-conducting material, so that, by the aid of a
handle, a contact strip can be brought into connection with
any one of them. The coils being arranged in series, the
movement of the handle in one direction adds resistance to the
field circuit, and in the opposite direction, cuts resistance out

FIGS. 170 AND 171.—FORMS OF FIELD RHEOSTAT.

of the circuit. Figs. 170 and 171 show different forms of *field
rheostats*, with wheel controlling handles. In some rheostats
the resistance wire is embedded in an enamel, which is caused
to adhere to a plate of cast iron. This gives a very compact
form of resistance; for, the intimate contact of the wire with
the iron plate, together with the large free surface of the plate,
enables the heat to be readily dissipated and prevents any
great elevation of temperature from being attained. Two of
such rheostats are shown in Fig. 172.

258. Compound-wound machines can be made to regulate
automatically, and do not require to have their E. M. F.

adjusted by the aid of a field rheostat. For this reason they are very extensively used in the operation of electric motors.

Series-wound machines are invariably used for operating arc lamps in series. Since the load they have to maintain is apt to be variable, such machines must possess the power of varying their E. M. F. within wide limits. Two methods are in use for maintaining constant the strength of current. That in most general use is to shift the position of the collecting brushes on the commutator so as to take off a higher or lower E. M. F. according as the load in the external circuit increases or decreases. The effect of this shifting will be evident from an inspection of Fig. 156; for, if the diameter of commutation be

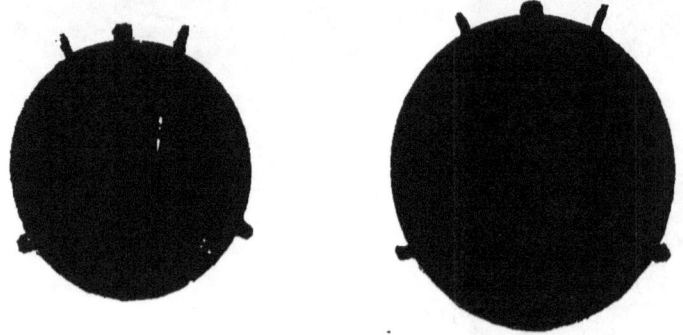

FIG. 172.—ENAMEL RHEOSTATS.

shifted to the right or left, the E. M. F. in some of the coils will be opposed to that in the remainder, the difference only being delivered at the brushes. In practice, the diameter of commutation would never reach the position of maximum E. M. F. represented in Fig. 156, and might, on the other hand, rotate through a sufficiently large angle to produce only a small fraction of the total E. M. F.

259. In all cases where the brushes are shifted through a considerable range over the commutator, care has to be taken to avoid the sparking that is likely to ensue if a certain balance is not maintained between the M. M. F. of the armature and the magnetic intensity in the air-gap. The fact that the current strength through the armature coils is practically constant at all loads, enables this balance to be effectually maintained, when once it has been reached at any load.

260. Series-wound arc-light generators have their armatures wound in two ways; namely, *closed-coil armatures*, and *open-coil armatures.* In the former, all the armature coils are constantly in the circuit, while in the latter, some of the coils are cut out of the circuit by the commutator, during a portion of the revolution. The ordinary continuous-current generator for producing constant pressure is, therefore, a closed-coil armature. Fig. 173 represents diagrammatically a form of open-coil armature winding. The three coils shown are connected to a com-

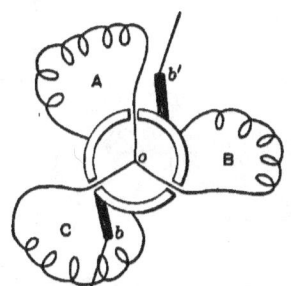

FIG. 173.—OPEN COIL-WINDING.

mon or neutral point o. In the position represented, the coil A, is disconnected from the circuits, the coils B and C, remaining in the circuit of the brushes b b'.

261. In closed-coil, series-wound, arc-light generators, the brushes are given a *forward lead; i. e.*, a lead in the direction of the rotation of the armature. The amount of this lead controls the E. M. F. produced between the brushes. It is essential, in order to prevent violent sparking, that the coil under commutation should be running through an intensity sufficient to nearly reverse the current in the commuted coil during the time of its short circuiting. Since the current strength in the field, and also in the armature, is maintained constant at all loads, it is necessary that the intensity of flux, through which the commuted coils run, should be uniform, or nearly uniform, at all loads and of the proper degree to effect current reversal. The

M. M. F. of the field magnet, is constant and the M. M F. of the armature is also constant, but the flux produced by the M. M. F. of the armature varies with the position of the brushes and the number of active turns that exist in that portion of the armature which is covered by the pole-piece, on each side of the diameter of commutation. The pole-pieces are usually so shaped that as the number of active turns in the armature covered by each pole increase; *i. e.*, as the load and E. M. F. of the machine increase, the trailing pole corners become more nearly saturated, and by their increasing reluctance check the tendency to increase the flux from the armature, so that an approximate balance between the field flux and the armature flux is main-

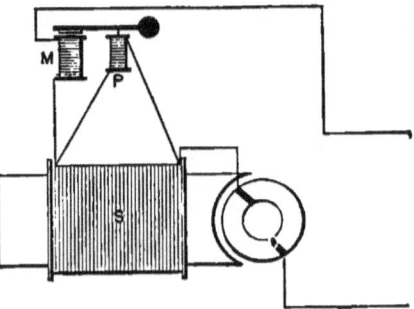

FIG. 174.—DIAGRAM OF AUTOMATIC REGULATOR CONNECTIONS.

tained at all loads. The armature flux always opposes the field flux at the diameter of commutation. The magnetic circuit, therefore, has to be so designed that the armature flux shall never quite neutralize the field flux at this point, but shall always leave a small residual field flux for the purpose of obtaining sparkless commutation.

262. The other method, which is employed for maintaining the current strength constant, introduces a variable shunt around the terminals of the field coil, in such a manner that when the current through the circuit becomes excessive, the shunt is lowered in resistance, and diverts a sufficiently large amount of current from the field magnets to lower their M. M F. to the required value. In order, however, to avoid the necessity for making this regulation by hand, it may be effected automatically as follows : namely, an electromagnet, situated in the main circuit, is caused by the attraction of its armature, on an increase in the main current strength, to bring pressure upon a pile of carbon discs. This pile of discs offers a certain resistance to the passage of a current, the resistance of the pile diminishing as the pressure upon it increases. The pile is placed as a shunt around the field magnet, so as to divert from the magnet a portion of the main current strength. When the attraction on the armature of the electromagnet increases the pressure on the pile, the resistance of the shunt path is diminished, and less current flows through the field magnets, as represented in Fig. 174, where S, is the series winding, shunted by the carbon pile P, and M, is the controlling magnet inserted in the main circuit.

263. Both the above methods are capable of compensating not only for variations in the resistance, or C. E. M. F. of the circuit, but also for variations in the speed of driving. In this respect the compensation is more nearly complete than that of constant pressure machines; for, compound-wound genertors can maintain a constant pressure under variations of load, but not under variations of speed.

CHAPTER XXI.

COMBINATIONS OF DYNAMOS IN SERIES OR IN PARALLEL.

264. When a system of electric conductors is supplied from a central station, it is evident, that if the load on the system was constant, a single large generator unit would be the simplest and cheapest source of electric supply, except, perhaps, on the score of reserve, in case of accidental breakdown. In practice, however, the load is never constant, and, therefore, the capacity of the generating unit is always considerably less than the total activity that has to be supplied at the busiest time. Moreover, engines and generators are necessarily so constructed, that while they may be comparatively very efficient when working at full load, they are far less efficient when working at a small fraction of their load, so that it is desirable to maintain such units as are in use, at full load under all circumstances. This consideration of wasted power, in operating large units at light loads, applies with less force to plants operated by water power, but, even in this case, it is usually found uneconomical to operate a large generator, for many hours of a day, when a smaller one would be quite competent to supply the load.

265. The generating units in a central station are, therefore, so arranged that they may be individually called upon at any time to add their activity to the output of the station. Electrically, these generators must be connected either in separate circuits, or in series or in parallel in the same circuit.

The method of connecting dynamos in series, so far as continuous-current circuits are concerned, is only employed for arc lamps operated in series. When a great number of arc lamps have to be supplied over a given district, they are usually arranged in different circuits, each circuit containing approximately the same number of lamps. Each such circuit is then connected, as a full load, to a single arc-light generator.

When, however, owing to some failure of continuity in a circuit, it is found impossible to operate two circuits independently, it is sometimes desirable to connect the two circuits together at some point outside the station, and to operate the increased load of lamps by two or more dynamos connected in series.

266. Generators are also connected in series when it is desired to employ, on the external circuits, the sum of the pressures of those generators. For example, in cases of the transmission of power to considerable distances, a high pressure in the conducting circuit is economically necessary. Whenever this pressure is greater than that which can be readily obtained from a single continuous-current generator, it is possible to connect two or more generators in series, so as to obtain the sum of their pressures. Thus, five generators, each supplying 500 volts pressure, will, when connected in series, supply a total pressure of 2,500 volts. The plan is rarely followed.

267. As a modification of the above plan, which is rarely adopted, five-wire, and three-wire systems, employing respectively four and two generators in series, are in use. The five-wire system, although employed in Europe, has not found favor in the United States. The three-wire system, however, is extensively employed. In this system, two generators of equal voltage, say 125 volts, are connected in series so as to supply a total pressure of 250 volts. Such a pressure is capable of operating incandescent lamps in series of two. To enable single lamps, however, to be operated independently, a third or *neutral wire* is carried through the system from the common connection point of the two generators, and the distribution of lamps, on the two sides of the system, is so arranged that the equalizing current, passing through the neutral wire, is small, and nearly as many lamps are operated at any one time on the positive, as on the negative side of the system. A pair of generators connected for three-wire service, therefore, constitutes a generating unit in a three-wire central station.

268. Series-generators are never, in practice, connected in parallel. Shunt-wound and compound-wound machines are capable of being connected in parallel, and most central stations arrange the generators in such a manner that they may be connected to, or disconnected from, the mains according to the requirements of the load.

269. Central stations, supplying incandescent lamps in parallel, usually employ shunt-wound generators, for the reason that the efficient and economic operation of the lamps requires a nearly uniform pressure at all lamp terminals.

Not only does the uniformity in the amount of illumination from an incandescent lamp depend upon the uniformity of the pressure supplied at its terminals, very small variations in the pressure markedly varying the intensity of light, but also such variations of pressure materially affect the life of the lamp. Thus a 50-watt, 16 candle-power, incandescent lamp, intended to be operated at a pressure of 115 volts, would have its probable life reduced by about 15 per cent., if operated steadily at 116 volts, and reduced by about 30 per cent., if operated steadily at 117 volts pressure. For this reason the pressure in the street mains supplying the lamps requires constant careful attention. Since it would be impossible to obtain at the mains a sufficient uniformity of pressure, under all conditions of load, by compound winding, and hand regulation would still be required, there is an advantage in dispensing altogether with compound winding, and resorting to hand regulation, with shunt winding, for the entire adjustment.

270. When two or more generators are connected in parallel, it becomes necessary that the electromotive forces they supply shall be equal, within certain limits. If, for example, two generators are connected in parallel, each working at half load, then if the drop of pressure in each generator armature at full load is two per cent. of its total E. M. F., it is evident that it is only necessary to increase the pressure of one generator two per cent. above that of the other, in order that the pressure at the brushes of the first shall be equal to the E. M. F. generated in the armature of the second. Under these circumstances no current will flow through the armature of the second machine, and all the load will be thrown on the first machine. If the E. M. F. of the first machine be still further raised, the pressure at its brushes will be greater than the E. M. F. in the armature in the second, and a current will pass through the second armature in a direction opposite to that which it tends to produce, and, therefore, in a direction tending to rotate the second generator as a motor. In other words, the control of pressure between the two machines must be within closer limits than two per cent. Early in the history of central station practice, difficulties were experienced in controlling the pressure of multiple-connected dynamos within limits necessary to avoid this unequalizing action, but at the present time, the governing of the engines and the control of the field magnets are so reliable, that this difficulty has practically disappeared. It is important to remember, however, that the larger the generator unit employed, and the smaller the drop in pressure taking place at full load through its armature, the narrower is the limit of speed or regulation, in which independent units will equalize their load, although as a counteracting tendency, the larger will be the amount of power which, in case of disequalizing, will be thrown upon the leading machine tending to check its acceleration.

271. Compound-wound generators are almost invariably employed for supplying electric currents to street railway systems. This is principally for the reason that the load in a street railway system is necessarily liable to sudden and marked fluctuations, and these fluctuations would be liable to produce marked variations in the pressure at the generator terminals, if the machines were merely shunt wound. Such generators are operated in parallel units. Here, as in the case of shunt-wound machines, it is necessary that the E. M. F. generated by each machine should be nearly the same, in order that the load should be equally distributed; but instability of control is greater in the case of compound-wound machines than in the case of shunt machines, for the reason that when one of a number of parallel-connected shunt-wound machines accelerates, and thereby rises in E. M. F., so as to assume an undue share of the load, the drop in the armature thereby increases, and tends to diminish the irregularity, so that not only does the greater load tend to retard the engine connected to the leading machine, but also the drop in its armature aids in equalizing the distribution.

In the case of compound-wound machines in parallel, any acceleration tends, as before, to increase the E. M. F. of the generator and, therefore, its share of the load, but the series coil of the compound winding being excited by the additional load, tends to increase the output of the machine, and, therefore, the governing of the engine has to be entirely depended on to prevent disequalization. Of recent years, however, the plan has been widely adopted of employing an *equalizing bar* between compound-wound generating units operated in par-

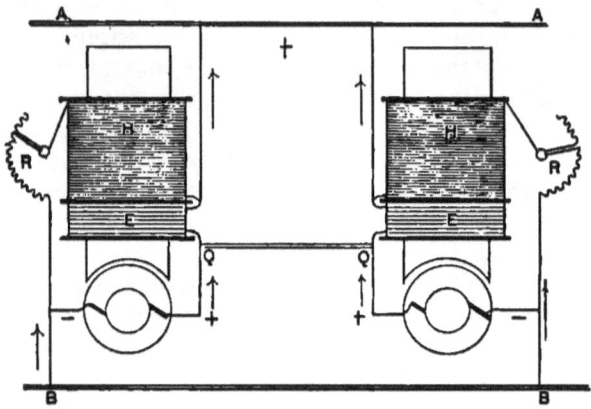

FIG. 175.—PARALLEL CONNECTION OF COMPOUND-WOUND GENERATORS.

allel. The connections of an equalizing bar are shown in Fig. 175. Here the two compound-wound generators are connected to the positive and negative *omnibus bars*, or *bus bars*, as they are generally termed, AA and BB, while the series coils are connected together in parallel by the equalizing bar QQ. It is evident that the equalizing bar connects all series coils of the different dynamos in parallel, so that any excess of current, supplied by the armature of one machine, must necessarily excite all the generators to the same extent.

272. When a number of compound-wound generators are running in parallel, and the load increases, so that it is desired to add another unit to the generating battery of dynamos, the engine connected with the new unit is brought up to speed, and the shunt field excited. This brings the E. M. F. of the machine up to nearly 500 volts. Its series winding is then connected in parallel with the series winding of the neighboring machines, by the switch on the equalizing bar, so that its excitation is then equal to that of all the other machines. The E. M. F. of the machine is then brought up slightly in excess of the station pressure by the aid of the field rheostat, and, as soon as this is accomplished, the main armature switch is closed, thus connecting the armature with the bus bars. The load of the machine is finally adjusted by increasing the shunt excitation, with the aid of the rheostat, until the ammeter connected with the machine shows that its load is approximately equal to that of the neighboring generators. The same steps are taken in reverse order to remove a generator from the circuit.

273. Fig. 176 is a diagram of a street-railway switchboard for two generators. It is customary, both for convenience and simplicity, to erect switchboards in panels, one for each generating unit, so that each panel controls a separate unit, and is in immediate connection with its neighbors. In the figure, the two panels are designated by dotted lines, the one on the left, active, and the one on the right, out of use. On each panel there are two main switches, P and N, for the positive and negative armature terminals. A smaller switch, not shown, is usually located on the right of each panel, and is for lighting up the station lamps from any panel and its connected machines, at will. R, is a shunt rheostat, placed at the back of the panel, with its handle extending through to the front, and S, is a small switch for opening and closing the shunt circuit of the field coils through the rheostat, R. A, is the generator ammeter, brought into use by the switches P and N, and T, is the automatic circuit-breaker for the panel. This electromagnetic circuit-breaker, opens the circuit of the machine when the current strength, owing to a short circuit or other abnormal condition, becomes dangerously great, thereby relieving the generator of the strain. The switch connected to the equalizing bar E is not placed in this instance, on the panel, but is mounted close to the generator with the object of diminishing the amount of copper conductor required. Each panel is also provided with a voltmeter connection and lightning arrester, which have been omitted here for the sake of simplicity.

274. The operations for introducing a unit into the battery of generators in this case, is as follows: the generator is brought up to speed, the equalizing switch is closed, thus connecting the series coils of the machines in parallel with the machines in use. The positive main switch P, is next closed, connecting

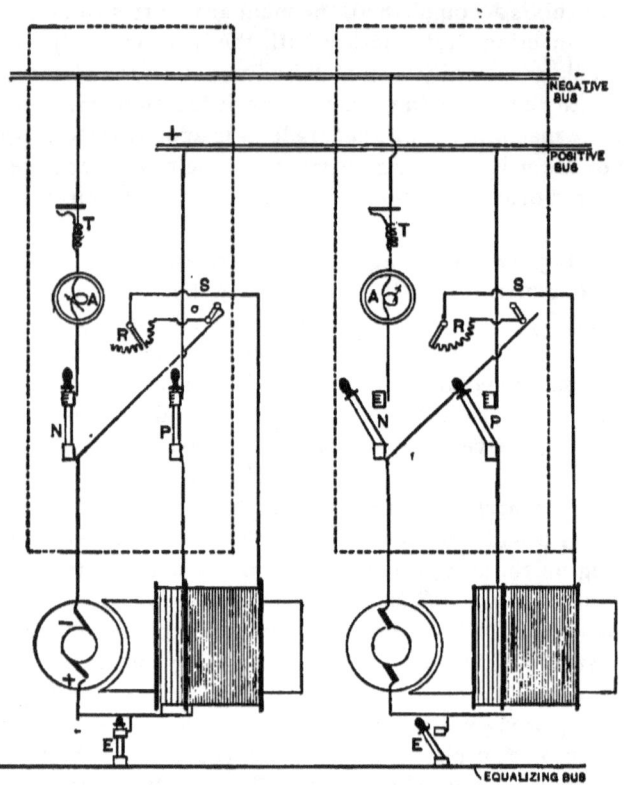

FIG. 176.—DIAGRAM OF SWITCHBOARD CONNECTIONS FOR TWO COMPOUND-WOUND GENERATORS.

one side of the armature to ground and to *return track feeders*. The field switch S, is next closed, and the E. M. F. of the machine brought up to slightly above station pressure by the aid of the rheostat R; finally, the negative main switch N, is closed, throwing the armature into the battery, and the load is adjusted by the rheostat R, in accordance with the indications of the ammeter A.

275. Another arrangement for railway switchboards consists in mounting the three switches, in close proximity to each other and attaching a single handle to the three blades, so that the three connections may be made or broken by a single operation.

When the railway mains are connected with the station by several feeders, it is customary to add another section to the switchboard where switches and ammeters are provided for handling the various feeders.

CHAPTER XXII.

DISC ARMATURES AND SINGLE-FIELD-COIL MACHINES.

276. Before leaving the subject of generators, it may be well to discuss a few types of generators that do not fall under the

FIG. 177.—DISC-ARMATURE GENERATOR.

types already discussed, and which are occasionally met with in practice.

These may be described as ;
(1.) Disc-armature machines.
(2.) Single-field-coil machines.

FIG. 178.—DISC ARMATURE.

(3.) Unipolar machines, or commutatorless continuous-current machines.

277. Generators employing disc armatures are frequently used in Europe, and although they are very seldom employed in the United States, yet it is proper to describe them as being types of machines capable of efficient use. In one form of disc-armature generator, the armature is devoid of iron, and is built of conducting spokes like a wheel, which revolves in a vertical plane between opposite field-magnet poles. Such a

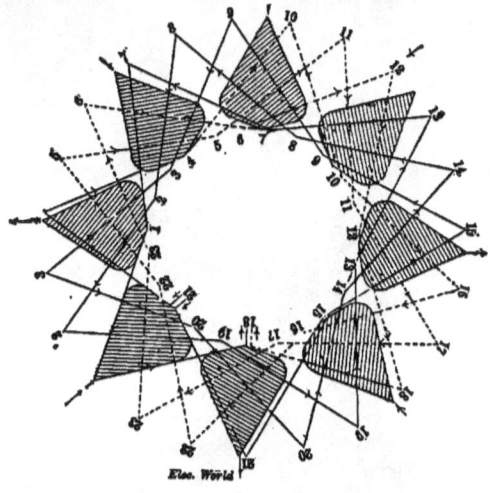

FIG. 179.—DIAGRAM OF DISC-ARMATURE WINDING.

disc-armature machine is shown in Fig. 177. It is to be observed that the entire machine is practically encased in iron, and is provided with three windows on the vertical face; through these windows the brushes, BB, rest on the commutator which is placed on the periphery of the disc, resembling in this respect the generator in Fig. 103. The armature of this machine is shown in Fig. 178 mounted on a suitable support. The radial spokes are of soft iron, and are connected into loops by the copper strips leading to the commutator segments on the periphery. The object of employing iron spokes is to diminish the reluctance of the air-gap. The field poles face each other, being separated by the disc armature, which revolves between them. Such an armature is evidently capable of being operated at an abnormally high temperature without danger, being constructed of practically fireproof materials. The electric connections of an octopolar machine are represented diagrammatically in Fig. 179. The brushes, it will be observed, are applied at the centres of any adjacent

FIG. 180.—DISC-ARMATURE GENERATOR.

pair of poles. Another form of the machine is represented in Fig. 180.

278. An example of a single-field-coil multipolar dynamo is shown in Fig. 181. This is a quadripolar generator with four sets of brushes. The interior of the field frame, with its projecting pole-pieces and exciting coil, is shown in Fig. 182. It will be seen that the field frame is made in halves,

FIG. 181.—COMPOUND-WOUND GENERATOR WITH SINGLE FIELD COIL.

FIG. 182.—DETAILS OF MAGNET, SINGLE-FIELD-COIL GENERATOR.

between which are enclosed the armature and the single field magnetizing coil. Four projections N, N, and S, S, form the pole-pieces of the quadripolar field; that is to say, the magnetic

FIG. 183.—ARMATURE OF QUADRIPOLAR, SINGLE-FIELD-COIL MACHINE.

flux produced by the M. M. F. of the single coil $C\ C$, passes through the field frame into the two pole faces N and N, in parallel through the armature into the adjacent pole faces S, S, thus completing the circuit through the field frame. The drum-wound, toothed-core armature, is shown in Fig. 183.

CHAPTER XXIII.

COMMUTATORLESS CONTINUOUS-CURRENT GENERATORS.

279. *Commutatorless continuous-current dynamos* are sometimes called *unipolar dynamos*, although erroneously. It is impossible to produce a single magnetic pole in a magnet, since all magnetic flux is necessarily circuital, and must produce poles, both where it enters and where it leaves a magnet. The fact that these machines are capable of furnishing a continuous current without the aid of a commutator, at one time caused considerable study to be given to them in the hope of rendering them

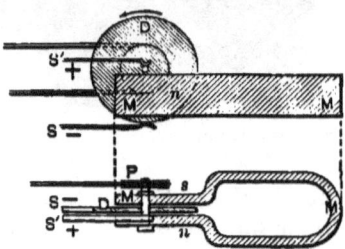

FIG. 184.—FARADAY DISC.

commercially practicable. The maximum E. M. F. which they have been constructed to produce, appears, however, to have been about six volts, and, consequently, they have practically fallen out of use, although they have been commercially employed for electroplating.

280. Fig. 184 represents what is known as a *Faraday disc*. This was, in fact, the earliest dynamo ever produced, and was of the so-called unipolar type; for here, a copper disc D, rotated, by mechanical force, about an axis parallel to the direction of the magnetic flux, supplied by a permanent horseshoe magnet MM, continuously cuts magnetic flux in the same direction, and, consequently, furnishes a continuous E. M. F. between the terminals S, S', without the use of a commutator.

281. The portion of the disc lying between the poles is caused to rotate in a nearly uniform magnetic flux, and with a velocity which depends upon the radius of the disc at the point considered, as well as on the angular speed of rotation. The direction of the E. M. F. induced will be radially downward from the axis to the periphery, and, if connection be secured between the axis as one terminal, and the rotating contact or brush as the other terminal, an E. M. F. will be continuously produced in that portion of the disc which lies beneath the poles; or, more strictly, in that portion of the disc which passes through the flux between them and around their edges. If, however, as in Fig. 185, the disc be completely covered by the pole faces, a

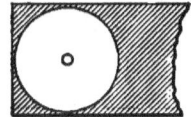 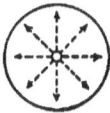

FIG. 185.—FARADAY DISC.

radial system of E. M. Fs. will be induced outward in the directions indicated by the arrows, or inward, if the direction of rotation be reversed. If no contacts are applied to the disc, these E. M. Fs. will supply no current, and will do no work. If brushes are applied at the axis, and at any or all parts of the periphery, the E. M. F. can be led off to the external circuit.

282. The value of the E. M. F. will depend upon the angular speed of rotation, the intensity of the magnetic flux, and the radius of the disc. The intensity of the magnetic flux can usually be made much greater by the use of a soft-iron disc instead of a copper disc, thereby practically reducing the reluctance of the magnetic circuit between the poles to that of two clearance films of air, since the reluctance of the iron disc will be negligibly small.

283. If we consider any small length of radius dl, Fig. 186, situated at a distance l, from the axis of the disc, the E. M. F.

generated in this element of the disc will be the product of the intensity, the length of the element, and its velocity across the flux. The element will be moving across the magnetic flux of uniform intensity, $\mathfrak{B}$ gausses, at a velocity $l\,\omega$ centimetres per second, where ω, is the angular velocity of the disc in radians per second. Consequently, the E. M. F. in this element will be:

$$de = l\,\omega\,.\,dr\,.\,\mathrm{B} \qquad \text{C. G. S. units of E. M. F.}$$

The total E. M. F. will be the sum of the elementary E. M. Fs. included in the radius taken from $l = o$, to $l = L$, the radius of the disc, or the integral of de, in the above equation between the limits $l = o$, and $l = L$. This integral is $\dfrac{L^2}{2}\,\omega\,\mathfrak{B} = e$. The E. M. F. from such a disc, therefore, increases as the

FIG. 186

square of the radius of the disc, directly as the speed, and directly as the uniform intensity of the magnetic flux. The same result can be obtained in a slightly different expression, since $\omega = 2\,\pi\,n$, where n, is the number of revolutions of the disc in a second, $e = \dfrac{L^2}{2}\,.\,2\,\pi\,n\,\mathfrak{B} = \pi\,L^2\,n\,\mathfrak{B} = S\,n\,\mathfrak{B}$ where S, is the active surface of the disc. This will also be true if the surface S, instead of extending over the entire face of the disc, extends only from the periphery to some intermediate radius. From this point of view the E. M. F. of the disc is equal to the product of the intensity in which it runs, the number of revolutions it makes per second, and its active surface in square centimetres. To reduce this E. M. F. to volts, we have to divide by 100,000,000.

284. There are two recognized types of commutatorless continuous-current dynamos; namely, the *disc type* and the *cylinder type*. The outlines of a particular form of the disc type are represented in Fig. 187. Here the shaft SS, usually horizontal, carries a concentric, perpendicular disc of copper or iron, rotating in a vertical plane, in the ring-shaped magnetic frame, in a circular groove, through the flux produced by two coils of wire. The general direction of the magnetic flux, through the field frame and disc, is represented by the curved arrows. It will be observed that the magnetic flux will be uniformly distributed so as to pass through the rotating disc at right angles. Brushes rest on the periphery, and on the shaft, of the disc. Inasmuch as the E. M. F. in the disc is radially directed at all points, the brushes for carrying off the current may be as numerous as is desired. These brushes are

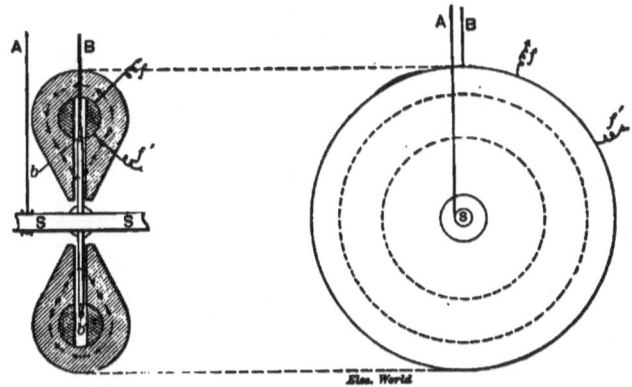

FIG. 187.—DISC TYPE OF COMMUTATORLESS DIRECT-CURRENT GENERATOR.

marked b, b, in the figure. A and B, are the main terminals of the machine, and f, f', the field terminals.

285. If we suppose that the intensity $\mathfrak{B}$, is 12,000 gausses, that the radius of the disc is 1 foot, or 30.48 centimetres, that the active surface on each side of the disc is 2,500 square centimetres, and that the speed of rotation is 2,400 revolutions per minute, or 40 revolutions per second, then the E. M. F. obtainable from the machine will be:

$$\frac{2,500 \times 40 \times 12,000}{100,000,000} = 12.0 \text{ volts.}$$

In order to produce an E. M. F. of say 140 volts, such as would be required for continuous-current central-station generators, it would be necessary either to connect a number of such machines in series, or to increase the diameter of the disc, or to increase the speed of rotation. It would, probably, be unsafe to run the disc at a peripheral speed exceeding 200 miles per hour, owing to the dangerously powerful mechanical stresses that would be developed in it by centrifugal force. This important mechanical consideration imposes a limit of speed of rotation and diameter of the disc, taken conjointly. By increasing, however, the active surface of the disc, and, at the same time, working at a safe peripheral velocity, it would

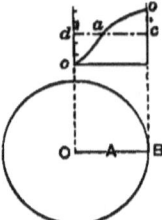

FIG. 188.—DIAGRAM SHOWING FLUX DENSITY THROUGH DISC ALONG A RADIUS.

be possible to construct large disc generators of this type for an E. M. F. of 100 or 150 volts.

286. It should be borne in mind that although such machines would be capable of producing continuous currents without the use of a commutator, yet the necessity of maintaining efficient rubbing contacts on the periphery of the rapidly-revolving disc introduces a difficulty and waste of power which has hitherto prevented the development of this system, and, probably, accounts for the fact that large machines of this type do not exist.

287. Irregularities in the distribution of magnetic flux over the surface of the disc may give rise to strong eddy currents and waste of power in the same. If the flux be variable along any radius of the disc OB, as represented in Fig. 188, so that the intensity $\mathfrak{B}$, is not uniform along these lines, this irregularity will not produce eddy currents in the disc unless the distribution is different along different radii. In other words, if the distribution of magnetic flux and intensity are symmetrical about the axis of rotation of the disc, the irregularities which exist will only alter the intensity of E. M. F. in different elements of a radius. In Fig. 188, the intensity, instead of being uniform from centre to edge, as indicated by the straight line *d a c*, increases toward the edge, following the line *o a b*.

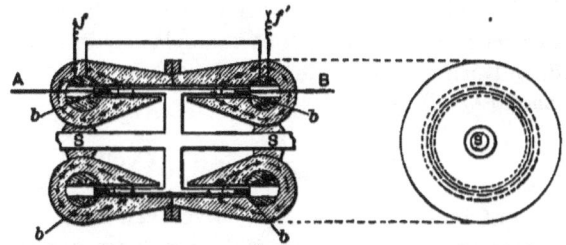

FIG. 189.—CYLINDER TYPE OF COMMUTATORLESS CONTINUOUS-CURRENT GENERATOR.

The formula for determining the E. M. F. of the disc is in such case rendered somewhat more complex.

288. If, however, the curve *o a b*, of flux intensity along different radii is different, so that the distribution of magnetic intensity is not symmetrical about the axis of rotation, then eddy currents will tend to form, the amount of power so wasted depending upon the amount of irregularity, the resistivity of the material in the disc, and the load on the machine.

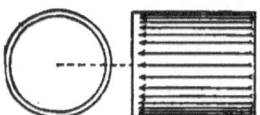

FIG. 190.—INDICATING DIRECTION OF E. M. F. INDUCED IN REVOLVING CYLINDER.

289. Fig. 189 represents the outlines of a particular form of the second, or cylindrical type of commutatorless continuous-current generator. Here a metallic conducting cylinder *cccc*, revolves concentrically upon the shaft *S S*, through the uniform magnetic flux, produced by the field frame surrounding it. Here, however, two sets of brushes *bb*, *b'b'*, have to be applied to the edges of the cylinder in order to supply the main terminals A and B. The terminals of the four circular coils constituting the field winding are shown at f, f'.

290. If the magnetic intensity produced by the field is uniform, the E. M. F. will be generated in lines along the surface of the cylinder parallel to its axis, as represented in Fig. 190. If v, be the peripheral velocity of the cylinder in centimetres per second, l, the length of the cylinder in centimetres, and $\mathfrak{B}$ the uniform intensity, in gausses, the E. M. F. generated by the machine will be:

$$e = \frac{v \, l \, \mathfrak{B}}{100{,}000{,}000} \quad \text{volts.}$$

Machines of the cylindrical type have been constructed and used for electrolytic apparatus, and give very powerful currents, as compared with ordinary generators of the same dimensions employing commutators. Unsatisfactory as these unipolar machines have so far proved, except in special cases, they are, nevertheless, the only dynamos which have yet been successfully constructed for furnishing continuous currents without the use of a commutator.

CHAPTER XXIV.

ELECTRO-DYNAMIC FORCE.

291. In discussing the magnetic flux surrounding an active conductor, we have observed in Par. 34, that it is distributed in concentric cylinders around the conductor, as shown in Figs. 27 and 28. It is evident that if a straight conducting

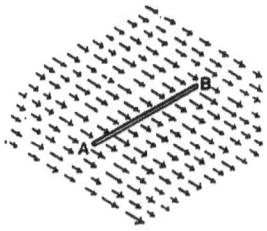

FIG. 191.—STRAIGHT CONDUCTOR IN UNIFORM MAGNETIC FLUX.

wire AB, say l cms. in length, as shown in Fig. 191, be situated in the uniform magnetic flux represented by the arrows, the flux will exert no mechanical influence upon the wire. If, however, the wire carries a uniform current in the direction from

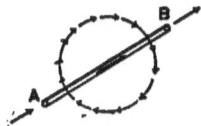

FIG. 192.—MAGNETIC FLUX SURROUNDING ACTIVE CONDUCTOR.

A to B, then, as is represented diagrammatically in Fig. 192, the system of concentric circular flux, indicated by a single circle of arrows, will be established around the wire, appearing clockwise to an observer looking from A, along the direction in which the current flows, and, as has already been pointed out, this circular magnetic flux will have an intensity proportional to the current strength.

292. If such a conductor be introduced into a uniform magnetic flux, as is represented in Fig. 193, it is evident that above the wire at C, the direction of the flux produced by the current is the same as that of the field, while below the wire at D, the direction of the flux from the current is opposite to that from the field. Consequently, the flux above the wire is denser,

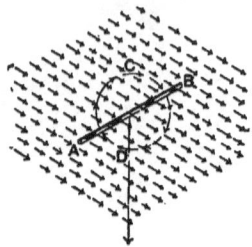

FIG. 193.—DIAGRAM SHOWING DIRECTION IN ELECTRO-DYNAMIC FORCE.

and that below the wire is weaker, or less dense, than that of the rest of the field. The effect of this dissymmetrical distribution of the flux density in the immediate neighborhood of the wire, is to produce a mechanical force exerted upon the substance of the wire, called the *electro-dynamic force*, tending to move it from the region of densest flux toward the region of weakest flux; or, in the case of Fig. 193, vertically down-

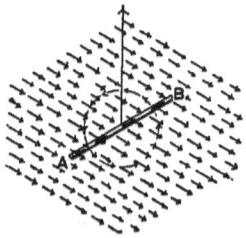

FIG. 194.—DIAGRAM SHOWING DIRECTION IN ELECTRO-DYNAMIC FORCE.

ward, as indicated by the large arrow. If, however, the direction of the current in the wire be reversed, as shown in Fig. 194, and that of the external field remain unchanged, the flux will be densest beneath the wire and weakest above it, so that the electro-dynamic force will now be exerted in the opposite direction, or vertically upward, as shown by the large arrows.

293. If the direction both of the current in the wire and the flux in the external field be reversed, the direction of the electro-dynamic force will not be changed, as is represented in Fig. 195, where the direction of the electro-dynamic force is downward as in Fig. 193, though the direction of the current and the direction of the magnetic field are both reversed.

294. A convenient rule for remembering the direction of the motion is known as *Fleming's hand rule*. It is, in general, the same as that already given for dynamos in Par. 81, except that in applying it, the left hand must be used instead of the right. For example, if the hand be held as in the rule for dynamos, if the *f*orefinger of the left hand shows the direction of the *f*lux, and the middle finger the direction of the cur-

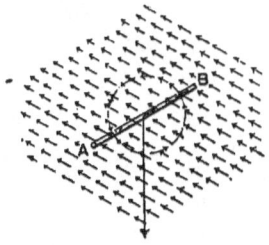

FIG. 195.—DIAGRAM SHOWING DIRECTION IN ELECTRO-DYNAMIC FORCE.

rent, then the thu*m*b will show the direction of the *m*otion. It must be remembered, that in applying Fleming's rule, the right hand is used for dynamos in determining the direction of the induced E. M. F., and the left hand for motors in determining the direction of motion.

295. We shall now determine the value of the electro-dynamic force in any given case, on the doctrine of the conservation of energy. To do this, we may consider the ideal apparatus, represented in Fig. 196, where a horizontal conductor EF, moves without friction against two vertical metallic uprights AB, and CD. This conductor is supported by a weightless thread, passing over two frictionless pulleys P, P, and bearing a weight W. If now a current enters the upright AB, and, passing through the sliding conductor EF, leaves the upright CD, at C, then, in accordance with the preceding principles, under the influence of the uniform magnetic flux passing horizontally across the bar in the direction of the arrows, an electro-dynamic force will act vertically downwards upon the rod. If this electro-dynamic force is sufficiently powerful to raise the weight W, it will evidently do work on such weight, as soon as it causes the bar to move. Let us suppose that it produces a steady velocity of the bar EF, of v cms. per second, in a downward direction. Then if f, be the

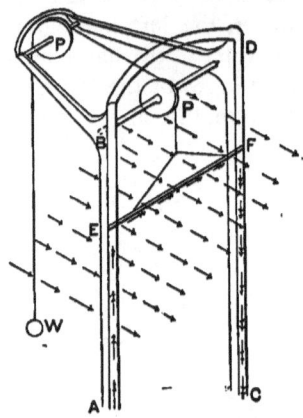

FIG. 196.—IDEAL ELECTRO-DYNAMIC MOTOR.

electro-dynamic force in dynes exerted on the bar, the activity exerted will be, vf centimetre-dynes-per-second, or ergs-per-second. Since 10,000,000 ergs make one joule, this will be an activity of

$$\frac{vf}{10,000,000}$$ joules-per-second, or watts.

This activity will be expended in raising the weight W, assuming the absence of friction. As in all cases of work expended, the requisite activity to perform such work must be drawn from some source, and in this case the source is the electric circuit.

296. When the bar of length l cms. moves with the velocity of v centimetres-per-second, through the uniform flux of density $\mathcal{B}$, it must generate an E. M. F. as stated in Par. 82, of $e = \mathcal{B}\,l\,v$, C. G. S. units, or

$$= \frac{\mathcal{B}\,l\,v}{100,000,000} \quad \text{volts.}$$

This E. M. F. is always directed against the current in the wire, and is, therefore, always a C. E. M. F. in the circuit. The current of i amperes passing through the rod will, therefore, do work upon this C. E. M. F. with an activity of

$$e\,i \quad \text{watts} = \frac{\mathcal{B}\,l\,v}{100,000,000}\,i \quad \text{watts.}$$

This activity must be equal to the activity exerted mechanically by the system, so that we have the equation,

$$\frac{v\,f}{10,000,000} = \frac{\mathcal{B}\,l\,v\,i}{100,000,000}$$

From which,

$$f = \frac{\mathcal{B}\,l\,i}{10} \quad \text{dynes.}$$

$\dfrac{i}{10}$ will be the number of C. G. S. units of current, since the C. G. S. unit of current is 10 amperes, so that the fundamental expression for the electro-dynamic force exerted on a straight wire, lying or moving at right angles across a uniform flux, is

$$f = \mathcal{B}\,l\,I \quad \text{dynes,}$$

where I, is expressed in C. G. S. units of current. Since the force of 981 dynes is, approximately, the force exerted by gravity upon one gramme, we have

$$f = \frac{\mathcal{B}\,l\,I}{981} \text{ or } \frac{\mathcal{B}\,l\,i}{9,810} \quad \text{grammes weight,}$$

and since 453.6 grammes make one pound, f, expressed in pounds weight will be

$$f = \frac{\mathcal{B}\,l\,i}{10 \times 981 \times 453.6} \quad \text{pounds weight.}$$

If, for example, the rod shown in Fig. 196 had a length of one metre, or 100 centimetres, and moved in the earth's flux whose horizontal component $= 0.2$ gauss, then if supplied with a uniform current of 1,000 amperes, it would exert a downward force of $0.2 \times 100 \times \dfrac{1,000}{10} = 2,000$ dynes; or approximately, 2 grammes weight.

297. We have heretofore considered the wire as lying at right angles to the flux through which it is moved. If, however, the wire AB, lies obliquely to the flux, at an angle β, as is represented in Fig. 197, then the *effective length of the wire*, or the projected length of AB, at right angles to the flux will be $a\,b$. In symbols this will be $l\sin\beta$, and the electro-dynamic force will be

$$f = \mathcal{B}\, l \sin\beta \, \frac{i}{10} \qquad \text{dynes.}$$

298. Although such a machine as is represented in Fig. 196 is capable of performing mechanical work, and might be, therefore, regarded as a form of electro-dynamic motor, yet all

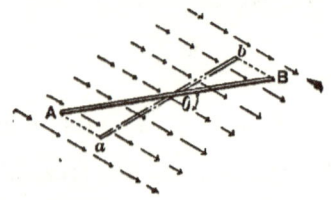

FIG. 197.—WIRE LYING OBLIQUE TO MAGNETIC FLUX.

practical electro-dynamic motors are operated by means of conducting loops, capable of rotating about an axis. We shall, therefore, now consider such forms of conductor.

299. If the rectangular loop $a'\ a''\ a'''\ a''''$, Fig. 198, placed in a horizontal plane, in a uniform magnetic flux, be capable of rotation about the axis oo, then if a current of i amperes be caused to flow through the loop in the direction $a'\ a''\ a'''\ a''''$, electro-dynamic forces will be set up, according to the preceding principles, upon the sides $a'\ a''$, and $a'''\ a''''$, but there will be no electro-dynamic force upon the remaining two sides. Under the influence of these electro-dynamic forces, the side $a'\ a''$, will tend to move upwards, and the side $a'''\ a''''$, downwards. The loop, therefore, if free to move, will rotate, and will occupy the successive positions b, c and d. At the last named position, the plane of the loop being vertical, although the electro-dynamic force will still exist, tending to move the the side $a'\ a''$, downwards, and the side $a'''\ a''''$, upwards, yet these forces can produce no motion, being in opposite directions and in the same plane as the axis; or, in other words, the loop considered as a rotatable system is at a dead point.

300. It is clear, from what has been already explained, that if the direction of the current in the loop had been reversed while the direction of the field flux remained the same ; or, if the direction of the field flux be reversed with the direction of current remaining the same, that the direction of the electro-dynamic forces would have been changed, tending to move the side $a'\ a''$, upwards and the side $a'''\ a''''$, downwards, so that the loop would have rotated in the opposite direction until it reached the vertical plane. Consequently, when a loop, lying

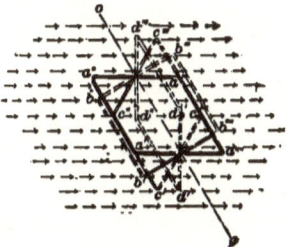

FIG. 198.—LOOP OF ACTIVE CONDUCTOR IN MAGNETIC FLUX.

in the plane of the magnetic flux, receives an electric current it tends to rotate, and, if free, will rotate until it stands at right angles to the magnetic flux.

301. An inspection of the figure will show that when the loop is in the plane of magnetic flux, that is to say, when the rotary electro-dynamic force is a maximum, the loop contains no magnetic flux passing through it, while when the loop is in the vertical position, and the rotary power of the electro-dynamic force is zero, it has the maximum amount of flux passing through it. The effect of the electro-dynamic force, therefore, has been to move the conducting loop out of the position in which no flux passes through it, into the position in which the maximum possible amount of flux passes through it, under the given conditions.

302. When an active conductor is bent in the form of a loop, such, for example, as is shown in Fig. 199, all the flux produced by the loop will thread or pass through the loop in the same direction, and this direction will depend upon the direction of the current around the loop. If, for example, we consider the loop $a^1\ a^2\ a^3\ a^4$, independently of the magnetic flux into which it is introduced, and send a current of i amperes, in the same direction as before around the loop, the general distribution of the flux around the sides of the loop is represented

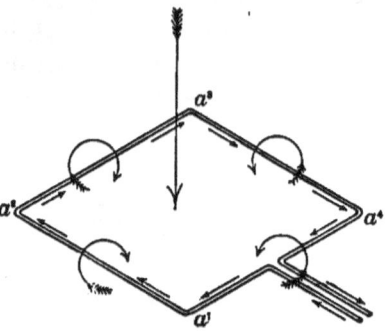

FIG. 199.—DIAGRAM SHOWING COINCIDENCE IN DIRECTION OF FLUX PATHS AROUND A LOOP OF ACTIVE CONDUCTOR.

by the circular arrows, from which it will be seen that all the flux passes downward through the loop as represented by the large arrow. If this loop be now introduced into the external magnetic flux, as shown in Fig. 192, it will tend to rotate, until the external magnetic flux passes through it in the same direction as the flux produced by its own current. Generally, therefore, it may be stated that when an active conducting loop is brought into a magnetic field, the electro-dynamic force tends to move the loop until its flux coincides in direction with that of the field.

303. During the rotation of the loop as shown in Fig. 198 from the position a, to the position d, the loop will embrace a certain amount of flux, say Φ webers, from the external field. In other words, in the position d, the loop holds Φ webers more flux than in the position a. If the current i amperes, passing through the loop be uniform during the rotation, then it can readily be shown that the amount of work performed by the loop during this motion is,

$$W = \frac{i \Phi}{10} \text{ ergs,}$$

but this motion comprises only one quarter of a complete revolution. At the same rate the work done in one revolution would be,

$$\frac{4 i \Phi}{10} \text{ ergs} = \frac{4 i \Phi}{10 \times 10,000,000} \text{ joules.}$$

304. In a bipolar motor with a drum-wound armature on which there are w wires, counted once completely around the periphery, or $\frac{w}{2}$ loops over the surface, there will be $\frac{w}{2}$ times as much work performed in one revolution as though a single loop existed on the surface; the work-per-revolution will, therefore, be

$$\frac{4 i \Phi}{100,000,000} \cdot \frac{w}{2} \text{ joules.}$$

If now the motor makes n revolutions per second, the work performed will be n times this number of joules in a second, or

$$\frac{4 i \Phi n}{100,000,000} \cdot \frac{w}{2} \text{ watts.} = \frac{2 i \Phi n w}{100,000,000} \text{ watts.}$$

Then, as will be shown hereafter, the current supplied at the brushes of the motor will be $I = 2 i$ amperes, if i, be the current through each loop, so that the activity absorbed by the motor will be,

$$\frac{I \Phi n w}{100,000,000} \text{ watts.}$$

We know that the E. M. F. of a rotating armature is

$$e = \frac{\Phi n w}{100,000,000} \text{ volts (see par. 132),}$$

so that we have simply, that the activity absorbed by the motor armature available for mechanical work is $e I$ watts, and this must be true under all conditions, in every motor.

When an E. M. F. of E volts acts in the same direction as a current I amperes; i. e., drives the current, it does work on the current with an activity of $E I$ watts, the activity being expended by the source of E. M. F. On the other hand, when an E. M. F. of E volts acts in the opposite direction to a current of I amperes, and therefore opposes it, or is a C. E. M. F. to the current, the current does work on the C. E. M. F. with an activity of $E\,I$ watts, and this activity appears at the source of C. E. M. F. If the C. E. M. F. be merely apparent in a conductor containing a resistance R ohms, as a drop $I\,R$ volts, the activity $E\,I = I^2\,R$, and is expended in the resistance as heat. If the C. E. M. F. be caused by electro-magnetic induction, as in a revolving motor armature, the activity $E\,I$, is expended in mechanical work, including frictions of every kind.

CHAPTER XXV.

MOTOR TORQUE.

305. We now proceed to determine the values of the rotary effort of a loop at different positions around the axis. This rotary effort is called the *torque*. Torque may be defined as the moment of a force about an axis of rotation. The torque is measured by the product of a force and the radius at which it acts. Thus, if in Fig. 200, a weight of P, pounds, be suspended from the pulley Y, and, therefore, acts at a radius l feet, the torque exerted by the weight about the axis will be $P\,l$ pounds-feet. If P, be expressed in grammes, and l, in centimetres, the torque will be expressed in gramme-centimetres; and if P, be in dynes and l, in centimetres, the torque

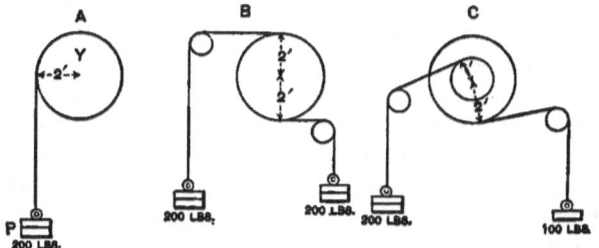

FIG. 200.—DIAGRAM ILLUSTRATING NATURE AND AMOUNT OF TORQUE.

will be expressed in dyne-centimetres. Thus, at A, Fig. 200, the torque about the axis of the pulley Y, is 400 pounds-feet. At B, it is 800 pounds-feet. At C, it is 400 pounds-feet.

As an example of the practical application of torque in electric motors, let us suppose that the pulley P, is attached to the armature shaft of a motor, and that the motor succeeds in raising the weight M, by the cord over the periphery of the pulley, then the motor will exert a torque at the pulley of $M\,l$ pounds-feet. Thus, if the pulley be 12 inches in diameter = 0.5 foot in radius, and the weight be 100 pounds, then if the thickness of the cord be neglected, the torque exerted by the motor will be 100 × 0.5 = 50 pounds-feet, about the shaft, at the pulley.

306. The work done by the torque which produces rotation through an angle β, expressed in radians, is the product of the torque and the angle. Thus, if the torque τ, rotates the system through unit angle about an axis, the torque does an amount of work $= \tau$. If the torque be expressed in pounds-feet, this amount of work will be in foot-pounds. If the torque be expressed in gm.-cms., the work will be expressed in cm.-gms., and finally, if the torque be expressed in dyne-cms. the work will be expressed in cm.-dynes, or ergs. Since there are 2π radians in one complete revolution, the amount of work done by a torque τ, in one complete revolution will be $2\pi\tau$ units of work. For example, the motor in the last paragraph, which produced a torque of 50 pounds-feet, would, in one revolution, do an amount of work represented by $50 \times 2\pi = 314.16$ foot-pounds. It is evident, in fact, that since the diameter of the pulley is one foot, one complete revolution will lift the weight M, through 3.1416 feet, and the work done in raising a 100-pound weight through this distance will be 314.16 foot-pounds. Similarly, if ω, expressed in radians per second, be the angular velocity produced by the torque, then the activity of this torque will be $\tau\omega$ units of work per second. For example, a motor making 1,200 revolutions per minute, or 20 revolutions per second, has an angular velocity of $20 \times 2\pi = 125.7$ radians per second. If the torque of this motor be 10,000 dyne-cms., the activity of this torque; $i.\ e.$, of the motor, will be $10,000 \times 125.7 = 1,257,000$ ergs per second = 0.1257 watt.

307. A torque must necessarily be independent of the radius at which it is measured. Thus, if a motor shaft is capable of lifting a pound weight at a radius of one foot; $i.\ e.$, of exerting a torque of one pound-foot, then it will evidently be capable of supporting half a pound at a radius of two feet, or one third of a pound at a radius of three feet, etc. In each case the torque will be the same; $i.\ e.$, one pound-foot.

308. The torque produced by a loop, situated in a uniform magnetic flux, varies with the angular position of the loop.

For example, returning to Fig. 198, the torque of the active loop is zero in the position d, and is a maximum in the position a. The electro-dynamic force exerted by the side $a'\,a''$ will be $\mathcal{B}\,l\,\dfrac{i}{10}$ dynes, and, if the radius at which this acts about the axis—i. e., half the length of the side $a'\,a''''$, be a cms., then torque exerted by this side will be $\dfrac{\mathcal{B}\,l\,i\,a}{10}$ dyne-cms. Similarly, the torque exerted in the same direction around the

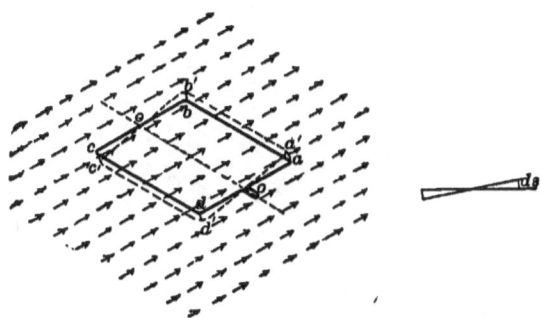

FIG. 201.—DIAGRAM SHOWING SMALL ANGULAR DISPLACEMENT ABOUT ITS AXIS, OF A LOOP IN UNIFORM MAGNETIC FLUX IN ITS PLANE.

axis by the side $a'''\,a'''$, will be also $\dfrac{\mathcal{B}\,l\,i\,a}{10}$ dyne-cms., so that the total torque around the axis will be $\dfrac{2\,\mathcal{B}\,l\,i\,a}{10}$ dyne-cms.

If the loop moves under the influence of this torque through a very small angle $d\beta$, the work done will be $\tau\,d\beta = \dfrac{2\,\mathcal{B}\,l\,i\,a}{10}\,d\beta$, but $a\,d\beta = ds$, the small arc moved through, as shown in Fig. 201, so that the work done will be $\dfrac{2\,\mathcal{B}\,l\,i\,ds}{10}$. The amount of flux linked with the loop during this small movement will be $2\,\mathcal{B}\,ds\,l = d\Phi$, so that the work done becomes $\dfrac{i}{10}\,d\Phi$, or I $d\Phi$ where I, stands for the current strength in C. G. S. units of ten amperes each. Consequently, in any small excursion of the loop, the work done will always be the product of the current strength and the increase of flux therewith enclosed. It is evident that the amount of flux which is brought within the loop by a given small excursion, varies with the position of the loop; that is to say, a small excursion through the arc ds, at the position represented both in plane and isometric projection, where the plane of the loop coincides with the direction of the flux, in Fig. 201, will introduce an amount of flux $= l \, ⓑ \, ds$. But the same small excursion in

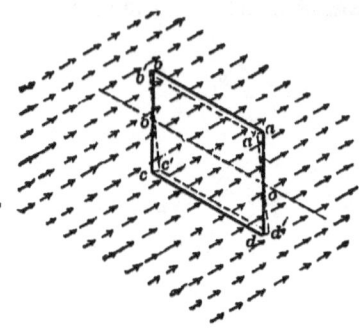

FIG. 202.—SMALL ANGULAR DISPLACEMENT OF A LOOP IN UNIFORM MAGNETIC FLUX PERPENDICULAR TO ITS PLANE.

the position represented in Fig. 202—*i. e.*, where the plane of the loop is perpendicular to the flux—will introduce practically no additional flux into the loop. At any intermediate position, it will be evident that the flux introduced by a small excursion of arc ds, will be $l \, ds \, ⓑ \cos \beta$, where β, is the angle included between the plane of the loop and the direction of magnetic flux. The torque exerted by the loop, therefore, varies as the cosine of the angle between the plane of the loop and the direction of the external flux.

309. Let us now consider the application of the foregoing principles to the simplest form of electro-magnetic motor. For this purpose we will consider a smooth-core armature A, Fig. 203, situated in a bipolar field. We will suppose that the total magnetic flux passing through the loop of the wire in the position shown, from the north pole N, to the south pole S, is Φ webers, and that a steady current of i amperes, is maintained through the loop of wire attached to the armature core. In the position of the loop as shown in Fig. 203, there will be no rotary electro-dynamic force exerted upon the wire, and the armature will be at a dead point. If, however, the armature be moved from this position into that shown in Fig. 204, so that it enters the magnetic flux, assumed to be uniformly distributed over the surface of the poles and armature core, then a rotary electro-dynamic force is set up on the wire, and com-

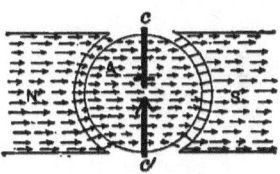

FIG. 203.—DRUM ARMATURE WITH SINGLE TURN OF ACTIVE CONDUCTORS AT DEAD POINT.

municated from the wire to the armature core on which it is secured. The torque being $\frac{i}{10} \cdot \frac{d\Phi}{d\beta}$ dyne-cms., where i, is the current strength in amperes, and $\frac{d\Phi}{d\beta}$ the rate at which flux enclosed by the loop is altered per unit angle of displacement. If, for example, the total flux $\Phi = 1$ megaweber, and the polar

FIG. 204.—ACTIVE CONDUCTOR ENTERING POLAR FLUX.

angle over which we assume that this flux is uniformly distributed is 120°, or $= \frac{2\pi}{3}$ radians, then the rate of emptying flux from the loop during its passage through the polar arc will be $\frac{1,000,000}{\frac{2\pi}{3}} = \frac{1,500,000}{\pi}$ webers-per-radian, and if the strength of current in the loop be maintained at 20 amperes, the torque exerted by the electro-dynamic forces around the armature shaft will be $\frac{20}{10} \times \frac{1,500,000}{\pi} = 955,000$ dyne-cms. Since a torque of 1 pound-foot = 13,550,000 dyne-cms., this torque would be represented by $\frac{955,000}{13,550,000}$ = 0.0705 pound-foot, or 0.0705 pound at one foot radius.

The armature will continue to move under this torque, if free to do so, until the position of Fig. 205 is reached, where

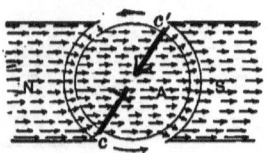

FIG. 205.—ACTIVE CONDUCTOR LEAVING POLAR FLUX.

it is evident that a still further displacement will not increase the amount of flux threaded through the loop.

The amount of work which will have been performed by the electro-dynamic forces during this angular displacement of 120° or $\frac{2\pi}{3}$ radians, will have been $\tau\beta = 955,000 \times \frac{2\pi}{3} = 2,000,000$ ergs, or, simply $\frac{i}{10} \Phi = \frac{20}{10} \times 1,000,000 = 2,000,000$ ergs = 0.2 joule.

310. The armature may continue by its momentum to move past the position of Fig. 205, to that of Fig. 206. As soon as it

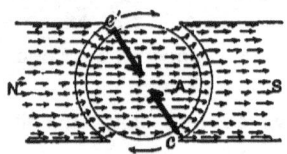

FIG. 206.—ACTIVE CONDUCTOR RE-ENTERING POLAR FLUX, AND ACTED ON BY OPPOSING ELECTRO-DYNAMIC FORCE.

reaches the latter position, a counter electro-dynamic force will be exerted upon it, tending to arrest and reverse its motion. Consequently, if the electro-dynamic force is to produce a continuous rotation, it is necessary that the direction of the current through the coil be reversed at this point; *i. e.*, commuted, or the direction of the field be reversed as soon as this point is reached. As it is not usually practicable to reverse the field, the direction of current through the coil is reversed by means of a commutator, so that when the position of Fig. 206 is reached, the current is passing through the wire in the opposite direction to that as shown by the arrow. Under these circumstances, the electro-dynamic force and torque continue in the same direction around the axis of the armature and expend another 0.2 joule upon the armature in its rotation to the original position shown in Fig. 203.

It is to be remembered that the representation of the flux in Figs. 203–206 is diagrammatic, since the flux in the entrefer is rarely uniform, never terminates abruptly at the polar edges, and is, moreover, affected by the flux produced around the active conductor.

311. The total amount of work done in one complete revolution of the armature upon a single turn of active conductor is, therefore, $\dfrac{2\,i\,\Phi}{10}$ ergs, or $\dfrac{2\,i\,\Phi}{100{,}000{,}000}$ joules.

If the load on the motor be small, so that the momentum of the armature can be depended upon to carry it past the dead-points which occur twice in each complete revolution, the armature will make, say n, revolutions per second, and the amount of work absorbed by the armature loop in this time will be $\dfrac{2\,i\,\Phi\,n}{100{,}000{,}000}$ joules in a second, or an activity of $\dfrac{2\,i\,\Phi\,n}{100{,}000{,}000} =$ watts.

The E. M. F. generated by the rotation of this loop through the magnetic field, by dynamo action, will be $\dfrac{\Phi\,n\,w}{100{,}000{,}000}$ volts, (Par. 132) where w, in this case is 2, since there are two conductors upon the surface of the armature, counting once completely around. The C. E. M. F. will, therefore, be $\dfrac{2\,\Phi\,n}{100{,}000{,}000}$ volts, and the activity of the electric current upon this C. E. M. F. will be $\dfrac{2\,i\,\Phi\,n}{100{,}000{,}000}$ watts, as above. Hence it appears that in this, as in every case, the torque and work produced by an electro-magnetic motor depends upon the C. E. M. F. it can exert as a dynamo.

312. Fig. 207 represents a Gramme-ring armature, carrying a single turn of conductor, situated in a bipolar field. If the total useful flux through the armature is Φ webers, as before, half of this amount will pass through the turn, or $\dfrac{\Phi}{2}$ webers, since the flux divides itself into two equal portions, as represented in the figure. It will be evident, as before, that starting at the position of Fig. 207, there will be no rotary electro-dynamic force exerted upon the loop, until it enters the flux,

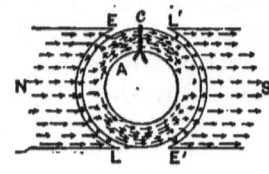

FIG. 207.—GRAMME-RING ARMATURE WITH SINGLE TURN OF ACTIVE CONDUCTOR AT DEAD POINT.

assumed to commence beneath the edge of the pole-piece, and the torque will then be uniform at the value $\dfrac{i}{10} \cdot \dfrac{d\Phi}{d\beta}$ dyne-centimetres, until the turn emerges from beneath the pole-piece at L. The work done in this passage will have been $\dfrac{i}{10} \cdot \dfrac{\Phi}{2}$ ergs, and this work will have been taken from the circuit, and, therefore, from the source of E. M. F. driving the current i, and will be liberated as mechanical work (including frictions). If, by the aid of the commutator, the direction of the current around the loop be reversed, the turn, when caused, either by momentum or by direct displacement, to enter the field at E, Fig. 208, will again receive a rotary electro-dynamic force whose torque is $\dfrac{\Phi i}{10} \cdot \dfrac{d}{d\beta}$ until the angle β, has been again passed, when the work performed will be $\dfrac{i}{10} \cdot \dfrac{\Phi}{2}$, ergs, as before. The total work done upon the armature in one revolution will, therefore, be $2 \times \dfrac{i}{10} \times \dfrac{\Phi}{2} = \dfrac{i\Phi}{10}$ ergs, and if the armature make n revolutions per second, the activity expended upon it will be $\dfrac{i\Phi n}{10}$ ergs per second $= \dfrac{i\Phi n}{100,000,000}$ watts · but considering the rotating armature in this case, as a dynamo armature, its E. M. F. will average $\dfrac{\Phi n}{100,000,000}$ volts, since there is only one turn of the wire upon its surface, and, consequently, the activity expended on the armature will be $e\,i = \dfrac{i\,\Phi\,n}{100,000,000}$ watts.

313. We have hitherto considered that the armature, whether of the Gramme-ring or drum type, possessed only a single

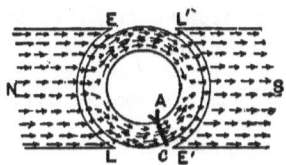

FIG. 208.—GRAMME-RING ARMATURE WITH SINGLE TURN OF ACTIVE CONDUCTOR.

turn. As a consequence the torque exerted by a constant current in the armature will vary between a certain maximum and zero, that is to say, the motor will possess dead-points. If, however, a number of turns be uniformly wound upon the armature, as in the dynamos already considered, it will be evident that the same number of turns will always be situated in the magnetic flux beneath the poles and in the air space beyond them, in all positions of the armature, and that, consequently, the torque exerted upon the armature will be constant when the magnetic flux and the current strength are constant. The torque exerted by the armature with w wires upon its surface, counted once completely around, will be $\dfrac{i}{10} \cdot \dfrac{\Phi w}{2\pi}$ dyne-cms., whether for a Gramme-ring or a drum armature, and this whether the armature be smooth-core or toothed-core.

That this is the case will be evident from the following consideration. The work done on a single wire in one complete revolution is $\dfrac{i\,\Phi}{10}$ ergs, and if there are w wires on the surface of the armature, the total work done by electro-dynamic forces in one revolution will be $\dfrac{i\,\Phi\,w}{10}$ ergs. But the work done by a torque τ dyne-cms. exerted through an angle of β radians is $\tau\beta$ cm.-dynes or ergs, and since one revolution is 2π radians, the work done by the torque will be $2\pi\tau$ ergs. Therefore,

$$2\pi\tau = \frac{i\,\Phi\,w}{10}, \text{ or } \tau = \frac{i\,\Phi\,w}{10} \cdot \frac{1}{2\pi} \text{ dyne-cms.}$$

For example, if a Gramme-ring armature has 200 turns of wire, counted once all round the surface, and the current strength supplied to the armature from the external circuit to the brushes is 50 amperes, while the total useful flux passing from one pole through the armature across to the other pole is 5,000,000 webers, or 5 megawebers, then the torque exerted by the armature under these conditions will be,

$$\frac{50}{10} \times \frac{500{,}000{,}000 \times 200}{2\pi} = 795{,}800{,}000 \text{ dyne-cms.} = \frac{795{,}800{,}000}{13{,}550{,}000}$$

pounds-feet $= 58.73$ pounds-feet.

314. The torque produced by multipolar continuous-current motors is independent of the number of poles, if the armature winding be of the *multiple-connected type; i. e.*, if there are as many complete circuits through the armature as there are poles in the field. In every such case, if Φ, be the useful flux in webers passing from one pole into the armature, i, the total current strength delivered to the armature in amperes, and w, the number of armature conductors counted once completely around its surface, the torque will be,

$$\frac{i\,\Phi\,w}{20\,\pi} \quad \text{centimetre-dynes, or}$$

$$\frac{i\,\Phi\,w}{20\,\pi \times 13{,}550{,}000} \quad \text{pounds-feet.}$$

If, however, the armature be series-connected, so that there are only two circuits through it, and there are p, poles in the field frame, the torque will be

$$\frac{p}{2} \cdot \frac{i\,\Phi\,w}{20\,\pi \times 13{,}550{,}000} \quad \text{pounds-feet.}$$

315. In a smooth-core armature, the electro-dynamic force, and, therefore, the torque, is exerted upon the active conductors, that is to say, the force which rotates the armature acts on the conductors which draw the armature around with them. Consequently, a necessity exists in this type of motor to attach the wires securely to the surface of the core in order to prevent mechanical displacement.

316. In a toothed-core armature, where the wires are so deeply embedded in the surface of the core as to be practically surrounded by iron, the electro-dynamic force or torque is exerted on the mass of the iron itself, and not on the wire. That is to say, the armature current magnetizes the core, and the magnetized core is then acted upon by the field flux. As soon as the iron of the armature core becomes nearly saturated by the flux passing through it, the electro-dynamic force will be exerted in a greater degree upon the embedded conductors, but, under ordinary conditions, the electro-dynamic force which they receive is comparatively small. A toothed-core armature, therefore, not only serves to protect its conductors from injury, since they are embedded in its mass, but also prevents their receiving severe electro-dynamic stresses. It is not surprising, therefore, that the tendency of modern dynamo construction is almost entirely in the direction of toothed-core armatures.

317. It might be supposed that the preceding rule for calculating the value of the torque in a motor, whether running or at rest, would only hold true where there existed a fairly uniform distribution of the field flux, such as would be the case where there was no marked armature reaction. Observations appear to show, however, that if we take into consideration the actual resultant useful flux which enters the armature from any pole, the torque will always be correctly given by the preceding rule, even when the armature reaction is very marked. That is to say if Φ, be the total useful flux passing through the armature from one field pole, the torque will be $\dfrac{i\,\Phi\,w}{20\,\pi}$ dyne-centimetres, no matter how much flux may be produced independently by the M. M. F. of the armature.

318. We have hitherto studied the fundamental rules for calculating the torque in the case of any continuous-current motor, whether bipolar or multipolar. It is well to observe that in practice the torque available from a motor at full load can be determined without reference to either the amount of useful flux passing through the armature, or to the amount of full-load current strength. For, if the full-load output of a motor be P watts, and the speed at which it runs be n revolutions per second, then the work done per second will be 10,000,000 P ergs. The angular velocity of the shaft will be $2 \pi n$ radians, and the torque, will, therefore be,

$$\tau = \frac{10,000,000\ P}{2 \pi n} \quad \text{dyne-centimetres.}$$

$$\tau = \frac{10,000,000}{13,550,000} \cdot \frac{P}{2 \pi n} \quad \text{pounds-feet.}$$

$$\tau = 0.1174 \frac{P}{n} \quad \text{pounds-feet.}$$

For example, if a motor gives six horse-power output at full load, and makes 600 revolutions per minute, required its torque.

Here the output, P, = 4,476 watts, the speed in revolutions per second $n = 10$, $\frac{P}{n} = 447.6$, and the torque exerted by the motor at full load will be,

$$\tau = 0.1174 \times 4,476 = 52.55 \text{ pounds-feet.}$$

If the amount of torque which the motor has to exert in order to start the load connected with it never exceeds the torque when running at full load, then the current which will be required to pass through the armature in order to start it will not exceed the full load current.

319. It is sometimes required to determine what amount of torque must be developed by a motor armature in order to operate a machine under given conditions. For example, if a machine has to be driven with an activity of ten horse-power, at a speed of 300 revolutions per minute, what will be the torque exerted by the motor running at 900 revolutions per minute, suitable countershafting being employed between machine and motor to maintain these speeds? If we employ the formula in the preceding paragraph, we find for the power $P = 10 \times 746 = 7,460$ watts. The speed $n = \frac{300}{60} = 5$ revolutions per second, so that the torque exerted at the shaft of the machine is $\tau = 0.1174 \dfrac{P}{n} = 0.1174 \times \dfrac{7,460}{5} = 175.1$ pounds-feet. The velocity-ratio of motor to machine is $\dfrac{900}{300} = 3$, so that the torque exerted by the motor, neglecting friction-torque in the countershafting will be $\dfrac{175.1}{3} = 58.37$ pounds-feet or 58.37 pounds at 1 foot radius.

Or, we might consider that the motor would, neglecting frictional waste of energy in countershafting, be exerting a power P of $10 \times 746 = 7,460$ watts at a speed of $n = \dfrac{900}{60} = 15$ revolutions per second. Its torque would then be, by the same formula, $\tau = 0.1174 \dfrac{P}{n} = \dfrac{0.1174 \times 7,460}{15} = 58.37$ pounds-feet.

320. In some cases it is necessary to determine the torque which must be exerted by a street-car motor at maximum load. It is not sufficient that the motor shall be able to exert a maximum activity of say 20 H. P. It is necessary that it shall be able to exert the given maximum torque at a definite maximum speed of rotation, and, therefore, the given maximum activity of 20 H. P. Otherwise, the motor might be of 40 H. P. capacity, and, yet by failing to exert the required torque, might be unable to start the car, or, in other words, the motor would have too high a speed.

For example, required the torque to be exerted by each of two single-reduction motors in order to start a car with 30″ wheels weighing 6 short tons light, and loaded with 100 passengers, up a ten per cent. grade, the gearing ratio of armature to car wheel being 3 to 1. Here 100 passengers may be taken as weighing 15,000 lbs. or 7½ short tons. The total weight of the car is therefore 27,000 lbs. The frictional pull required to start a car from rest on level rails, under average commercial conditions, is about 1.8 per cent. of the weight, or, in this case, 486 lbs. weight. The pull exerted against gravity is also 2,700 lbs., making the total pull 3,186 lbs. weight. The radius of the car wheel being $\dfrac{30}{24} = 1.25$ feet, the torque at the car wheel axle is $3,186 \times 1.25 = 3,983$ pounds-feet. The torque at the motor shafts is therefore $\frac{3,983}{3} = 1,328$ pounds-feet, and each motor must therefore exert $\frac{1,328}{2} = 664$ pounds-feet.

If the motors make 600 revolutions per minute or 10 revolutions per second, exerting this torque, their activity will be $664 \times 10 \times 2\pi \times 1.355 = 56,530$ watts, $= 56.53$ KW, and their combined activity 113.1 KW, neglecting gear frictions.

321. Considering the case of a motor armature in rotation, the speed of its rotation for a given E. M. F. applied to its armature terminals will depend upon three things : *viz.*,

(1.) The load imposed upon the armature, or the torque it has to exert.

(2.) The electric resistance of the armature in ohms.

(3.) Its *dynamo-power;* i. e., its power of producing C. E. M. F., or the number of volts it will produce per revolution per second.

If E, be the E. M. F. in volts applied to the armature terminals, τ, the torque, which the motor has to exert, including the torque of frictions, in megadyne-decimetres (dyne-cms. $\times$ 10^{-7}) r, the resistance of the motor armature in ohms, and e, the C. E. M. F. produced in volts per revolution per second of the armature. Then $n\,e$, will be the total C. E. M. F. $\frac{E-n\,e}{r}$ will be the current strength received by the armature according to Ohm's law. The activity of this current expended upon the C. E. M. F. will be their product, or $n\,e \times \frac{E-n\,e}{r}$ watts, and this must be equal to the total rate of working, or $2\pi n\tau$, $=$ consequently, $n\,e\left(\frac{E-n\,e}{r}\right) = 2\pi n\tau$ and

$$n = \frac{E}{e} - 2\pi \frac{r\,\tau}{e^2}$$ revolutions per second.

For example, if a motor armature, whose resistance is 2 ohms, has a uniformly excited field, which may be either of the bipolar or multipolar type, and is supplied with 500 volts at its terminals ; and if the C. E. M. F. it produces by revolution in the field is 40 volts per-revolution-per-second, then the speed at which the motor will rotate, when exerting a torque, including all frictions, of 100 pounds-feet (100 × 13,550,000 dyne-centimetres, = 135.5 megadyne-decimetres) will be

$$n = \frac{500}{40} - \frac{2\pi \times 2 \times 135.5}{1,600} = 12.5 - 1.06 = 11.44 \text{ revolutions-}$$

per-second.

322. It will be observed from the above formula that if either the torque be zero, or the resistance of the armature is zero, the speed of the motor will simply be $\frac{E}{e}$ revolutions-per-second. Or, in other words, that the armature will run at such a speed that its C. E. M. F. shall just equal the E. M. F. applied to the armature; *i. e.* without drop of pressure in the armature. If the torque could be made zero, the motor would do no work and would require no current to be supplied to it, so that no matter what the resistance of the armature might be, the drop in the armature would be zero. All motors necessarily have to exert some torque in order to overcome various frictions, but on light load their speed approximates to the value $\frac{E}{e}$ revolutions-per-second. If the resistance of the motor is very small, which is approximately true in the case of a large motor, the second term $\frac{2\pi r\tau}{e^2}$, in the formula, becomes small, and the diminution in speed due to load is, therefore, also small. In other words, the drop which takes place in the armature due to its resistance is correspondingly reduced, permitting the motor to maintain its speed and C. E. M. F. of rotation. Fig. 209 represents diagrammatically a motor armature revolving in a suitably excited magnetic field, and supplied from a pair of mains, *M*, *M*, with a steady pressure of 500 volts. The resistance of the armature is represented as being collected in the coil r, while the C. E. M. F. of the motor is indicated as opposing the passage of the current from the mains.

The drop in the resistance is represented as being 40 volts, while the C. E. M. F. is 500 − 40, or 460 volts.

323. The E. M. F. applied to the terminals of a motor armature, therefore, has to be met by an equal and opposite or

C. E. M. F. in the armature, which is composed of two parts, that due to rotation in the magnetic flux, or to dynamo-electric action, and that apparent C. E. M. F. which is entirely due to drop of pressure in the resistance of the armature, considered as an equivalent length of wire. The activity expended against the C. E. M. F. of rotation is activity expended in producing torque, and, therefore, almost all available for producing useful work, while the activity expended against the C. E. M. F. of drop is entirely expended in heating the wire. As the load on the motor is increased, the current

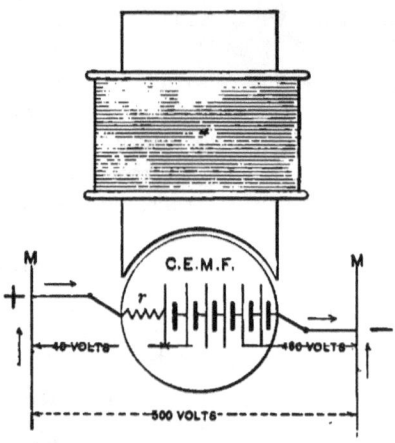

FIG. 209.—DIAGRAM REPRESENTING RESISTANCE AND C. E. M. F. IN A REVOLVING MOTOR-ARMATURE.

which must be supplied to the motor to overcome the additional load or torque increases the drop in the armature, and, therefore, diminishes the C. E. M. F. which has to be made up by rotation, and the speed falls, or tends to fall, in proportion.

324. When a motor armature is at rest, its C. E. M. F. of rotation is zero, and the C. E. M. F. which it can produce under these conditions must be entirely composed of drop of pressure. In other words, the current which will pass through it is limited entirely by the ohmic resistance of the circuit.

If i, be the current strength in amperes supplied to a motor armature at a pressure of E volts, the activity expended in the armature will be $E\,i$ watts. The activity expended in producing torque will be $n\,e\,i$ watts, so that disregarding mechanical and electro-magnetic frictions, the efficiency of the motor will be $\dfrac{n\,e\,i}{E\,i} = \dfrac{n\,e}{E}$, or simply the ratio of the C. E. M. F. of rotation to the impressed E. M. F. This is a maximum at no load; *i. e.*, when the motor does no work, and is zero when the motor is at rest.

The value of e, the volts-per-revolution-per-second, is in all cases of multiple-connected armatures equal to $\Phi\,w \times 10^{-8}$, where Φ, is the number of webers of flux passing usefully into the armature from any one pole, and w, is the number of turns of conductor counted once around its periphery.

325. The speed of a motor, therefore, varies, to the first approximation, inversely as the useful magnetic flux, and inversely as the number of armature conductors. A slow-speed motor, other things being equal, is a motor of large flux, or large number of turns, or both, and, as will afterward be shown, in order to decrease the speed at which the motor is running, it is only necessary to increase, by any suitable means, the useful flux passing through its armature.

326. Just as in the case of a generator armature, whose maximum output is obtained when the drop in its armature is equal to half its terminal E. M. F. (Par. 9), so in the case of the motor, the output is a maximum (neglecting frictions), when the drop in the armature is half the E. M. F. applied at the armature terminals, or, in symbols, when $n\,e = \dfrac{E}{2}$; the speed of the motor being then half its theoretical maximum speed, assuming no friction.

Similarly, just as it is impracticable to operate a generator of any size at its maximum theoretical output, since the activity expended within it would be so great as probably to destroy it, being equal to its external activity, so no motor of any size can be operated so as to give the maximum theoretical output of work, since the activity expended in heating the machine, being equal to its output, would, probably, cause its destruction.

CHAPTER XXVI.

EFFICIENCY OF MOTORS.

327. As in the case of generators, the *commercial efficiency* of the electric motor is the ratio of the output to the intake: that is,

$$\text{Efficiency} = \frac{\text{Output}}{\text{Intake}}.$$

Since the output must be equal to the intake after subtracting the loss taking place in the machine, the above may be expressed as follows:

$$\text{Efficiency} = \frac{\text{Intake} - \text{Losses}}{\text{Intake}}.$$

328. The losses which occur in a motor are of the same nature as those already pointed out in Par. 224, in connection with a generator. This is evident from the fact that a motor is but a generator in reversed action; so that any dynamo is capable of being operated, either as a generator or as a motor, according as the driving power is applied to it mechanically or electrically. There is this difference, however, between the two cases, that a very small dynamo-electric machine may be capable of acting as a motor, while it is not capable of acting as a dynamo, owing to the fact that it is not able, unaided, to excite its own field magnets, its residual magnetism being insufficient for this purpose. On this account, motors can be constructed of much smaller sizes than self-exciting generators.

329. If the losses which occur in a dynamo-electric machine, acting as a generator, have been determined, we can then closely estimate what these losses will be when the machine is operated as a motor, and, consequently, the efficiency of the machine as a motor can be arrived at.

330. There is this difference between a dynamo and a motor as regards the output; *viz.*, in the dynamo, the energy lost is derived from the driving source, while in the motor the energy lost is derived electrically from the circuit; but the output of a dynamo-electric machine is almost invariably determined by the electric activity in its armature circuit; that is to say, the armature is limited to a certain number of amperes received or delivered at a certain number of volts pressure, so that since this load is the output, when the machine is a generator, and the intake, when the machine is a motor, it is evident that after the losses as a motor have been subtracted, the mechanical output will be less than the electrical output which the machine produces as a generator.

331. For example, let us suppose that a certain machine, acting as a series-wound generator, is capable of delivering 10 amperes at a pressure of 100 volts, so that its output is 1 KW. Let us also suppose that when acting as a generator, a loss of 250 watts occurs, in friction, hysteresis, eddy currents and I^2R losses, both in the armature and in the field; then the mechanical intake of the machine will be 1,250 watts, and its commercial efficiency, $\frac{1,000}{1,250} = 0.8$, or 80 per cent. When, however, the machine is operated as a motor, the armature is limited to the same current strength of 10 amperes, and the pressure at the machine terminals can only be slightly in excess of the 100 volts previously delivered. Let us suppose that this is 110 volts. Then the intake of the machine will be 1,100 watts. Assuming the same losses as before; namely, 250 watts, the output would be only 850 watts, and the efficiency, therefore, $\frac{850}{1,100} = 0.772$, or about 2¾ per cent. less than in the preceding case. It is clear, therefore, that while the output of the machine was 1,000 watts when acting as a generator, it was limited to 850 watts when acting as a motor, assuming that the same limiting armature temperature and same liability to sparking were accepted in each case.

332. The difference above pointed out between the output of a machine acting as a generator and as a motor, diminishes with an increase in the size of the machine. Thus, while a 1-KW generator is usually only a 1-H. P. motor (or has an output of say 750 watts), a generator of 200 KW would, probably, be a motor of 185 H. P.; so that in the case of very large machines, the difference between the outputs in the two cases would be practically negligible.

333. The curve in the accompanying Fig. 210, approximately represents the efficiency which may be expected at full load

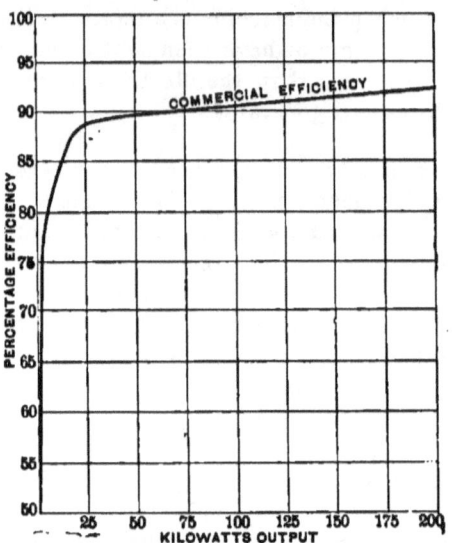

FIG. 210.—COMMERCIAL EFFICIENCY CURVE OF MOTORS AT FULL LOAD.

from motors of varying capacity up to 200 KW. This curve has been plotted from a number of actual observations with machines constructed in the United States.

334. It is to be remembered, however, that the full load efficiency of a motor is not always the criterion upon which its suitability for economically performing a given service is to be determined. It not infrequently happens that the character of the work which a motor has to perform is necessarily exceedingly variable, so that the average load might not be half the full load of the machine. Under such conditions, the *average efficiency* is of more importance than the *full-load efficiency*. Were the efficiency curve of all motors in relation to their load of the same general outline, the average efficiency would be, approximately, the same in all motors having the same full-load efficiency. As a matter of fact, however, the efficiency curves of different machines may be very different. Thus one machine may have its maximum efficiency at half load, and behave at full load, in regard to its efficiency, as though it were actually overloaded, while another machine, with the same full-load efficiency, may show a lower efficiency at half load. Obviously the first machine would be preferred for variable work, other things being equal.

335. Similar considerations apply to electric generators. The full-load efficiency is not in every case the ultimate criterion of economical delivery of work, but it generally happens that generators are installed in such a manner, and under such conditions, that a nearer approach to their full load is attained, so that ordinarily the shape of the efficiency curve of a generator is not of such great importance as that of a motor.

Fig. 211 represents the efficiency curves of two motors, each having a full-load efficiency of 78 per cent. One of these machines has an efficiency, at about two-thirds load, of 84 per cent., but at overloads is inefficient, while the other becomes more efficient at slight overloads.

336. In order to produce a motor of given full-load efficiency with comparatively small loss at moderate loads, and, therefore, a comparatively heavy loss at heavy loads, we may employ a *slow-speed motor*, or a motor which shall generate the necessary C. E. M. F. at a comparatively low speed. Such a machine will probably have a small loss in mechanical friction, because of its lower speed of revolution. It will, similarly, have, probably, a small loss in hysteresis and eddy currents for the same reason, but a slow speed motor will probably have a greater number of armature turns in order to compensate for the smaller rate of revolution, and the I^2R loss in the armature is, therefore, likely to be greater at full load. In such a machine, the loss at full load is principally due to I^2R; and, since this loss decreases rapidly with I, it will evidently have a small loss at moderate loads.

337. The speed at which a motor will run in performing a given amount of work varies considerably with different types of motors. For example, of two motors of 20 KW capacity, one may run at 400 revolutions-per-minute, and the other at 1,000 revolutions-per-minute. It is evident that the first machine will have two and a half times the full-load torque of the second. The lower speed is, however, generally speaking, only to be obtained at the expense of additional copper and iron ; that is to say, the cost of material in a slow-speed machine will, probably, be greater than the cost of material

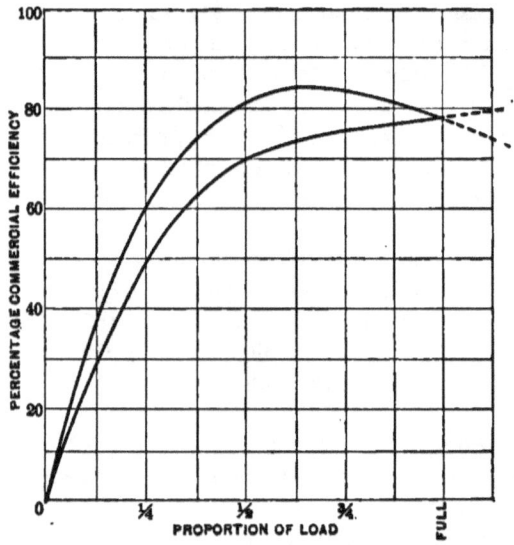

FIG. 211.—EFFICIENCY CURVES OF TWO DIFFERENT MOTORS HAVING SAME FULL-LOAD EFFICIENCY.

in a high-speed machine of the same output and relative excellence of design. It becomes, therefore, a question as to the relative commercial advantage of slow speed versus high speed in a motor.

338. Motors are generally installed to drive machinery either by belts or gears, and the belt speed or the gear speed of machinery is, in practice, a comparatively fixed quantity. If, therefore, the speed of the motor be greater than the speed of the main driving wheel of the machines with which the motor is connected, intermediate *reducing gear* or countershafting has to be installed. This adds to the expense of installation, not only in first cost, but also in maintenance, lubrication, and the continuous loss of power it introduces through friction. The result is, that up to a certain point, slow-speed motors are economically preferable, and the tendency of recent years has been toward the production of slower speed dynamo machinery. In comparing, therefore, the prices of two motors of equal output, the speed at which they run has to be taken into account, as well as the efficiency at which they will operate. It is to be remembered that any means in the design which will enable a motor to supply its output at a slower speed, are equivalent to means which will enable a motor of the higher speed to supply a greater output.

339. The weight of a motor is a matter of considerable importance in cases of *locomotors ; i. e.*, of *travelling motors*, as in the case of electric locomotives, street-car motors or launch motors, but in the case of *stationary motors*, their weight is of less consequence, since, after freight has been once paid for their shipment, no extra expense is entailed by reason of their increased mass when in operation. Indeed, weight is often a desirable quality for a motor to possess in order to ensure steadiness of driving, although undue weight in the armature is apt to produce frictional loss, and diminished efficiency.

340. In comparing the relative weights of motors, two criteria may be established; namely,

(1) In regard to torque, and (2) in regard to activity. In some cases, the work required from the motor is such that the pull or torque which must be given in reference to its weight is the main consideration; while in other cases it is not the torque, but the output per-pound of weight, which must be considered.

341. The torque-per-pound, in the case of street-car motors, where lightness is an important factor, has been increased to

133,000 centimetre-dynes per-ampere, per-kilogramme of weight; or, 0.0045 pound-foot per-ampere per-pound of total motor weight, exclusive of gears, so that a 500-volt street-car motor, weighing 223 pounds, and supplied with one ampere of current, would exert a torque of one pound-foot. In stationary motors, the torque is usually only 0.001 to 0.0015 pound-foot per-ampere per-pound of weight, or about four times less than with street-car motors. This is owing to the fact that cast iron is more largely employed in stationary motors, owing to its lesser cost.

The output per-pound of weight in motors varies from 5 watts per pound to 15 watts per pound, according to the size and speed of the motor.

342. We may now allude to the theoretical conditions which must be complied with in order to obtain the maximum amount of torque in a motor for a given mass of material. It must be carefully remembered, however, that these theoretical conditions require both modification and amplification, when applied to practice, so that the practical problem is the theoretical problem combined with the problem of mechanical construction.

343. The torque of a motor armature being $\dfrac{i \, \Phi \, w}{20 \, \pi}$ cm.-dynes, we require to make this expression a maximum for a given mass of copper wire in the armature and in the field magnets, neglecting at present all considerations of structural strength.

The torque-per-ampere will be $\dfrac{\Phi \, w}{20 \, \pi}$ cm.-dynes.

In order to make this a maximum, both Φ and w, should be as great as possible.

344. It is evident that if we simply desired a motor of powerful torque-per-ampere, regardless of its weight, we should employ as much useful iron as possible, so as to obtain as great a useful magnetic flux Φ, through the armature, as possible, and we should employ as many turns of wire upon the surface of the armature as could be obtained without making the armature reaction excessive, or without introducing too high a resistance, and too much expenditure of energy in the armature winding. Such a motor would essentially be a heavy motor, so that the requirements of a motor with powerful torque-per-ampere would simply be met by a motor of great useful weight, and this, indeed, would be obvious without any arithmetical reasoning.

345. When, however, the torque-per-ampere per-pound-of-weight has to be a maximum, the best means of attacking the problem is to consider a given total weight of copper and iron in the armature, and examine by what means this total weight can be most effectually employed for producing dynamo-power; *i. e.*, volts-per-revolution-per-second, and torque-per-ampere.

346. It will, in the first place, be obvious that a long magnetic circuit will not be consistent with these requirements, since, as we shorten the magnetic circuit, retaining the same mass of material, we make it wider, or of greater section, and so increase the total flux Φ. In the second place, the material of which the magnetic circuit is formed should have as small a reluctivity, and as powerful a flux density as possible, since this will increase the total flux without adding to the weight. For this reason soft cast steel is much to be preferred to cast iron.

347. Again, it will be evident that as we increase the number of turns on the armature, having determined upon a certain total mass of armature copper, or *armature winding space*, we increase, according to the formula, the torque-per-ampere. But, in occupying the given winding space with many turns instead of with few turns, we increase, for a given speed, the voltage of the armature. Thus, if a motor armature be intended to rotate at a speed of 10 revolutions per second, its E. M. F., other things being equal, will be 10 times as great, when we use 10 times as many wires upon its surface, and its torque-per-ampere will be also increased 10 times. A high E. M. F. motor is, therefore, necessarily a motor of high torque-per-ampere. A 500-volt armature would, therefore, in accordance with preceding principles, necessarily be a motor of greater torque-per-ampere than the same armature wound for 100 volts, although the torque at full load might be the same in each case, since the low-pressure armature might make up by increase of current what it lacked in torque-per-ampere.

348. Having selected a field frame with as short a magnetic circuit as is consistent with not excessive magnetic leakage, and with room for magnetizing coils, and having placed a large number of turns upon the armature surface, there remain several important detail considerations which should be taken into account to enable a high torque-per-ampere to be obtained.

349. In the first place, the reluctance in the magnetic circuit should be as small as possible in order to diminish the M. M. F. and the mass of magnetizing copper. With smooth-core armatures this would represent a small entrefer and a small winding space, whereas, to obtain many turns, we require a large entrefer and large winding space, so that with a smooth-core armature, a compromise is necessary at some point of maximum effect, depending upon a great variety of details. With toothed-core armatures, however, a large number of turns may be disposed upon the armature surface, yet the reluctance in the entrefer may be comparatively small. This consideration affords an additional argument in favor of toothed-core armatures for high torque.

350. In the second place, the number of poles in the field frame should be as great as possible. If we double the number of poles in the field frame, retaining the same armature, and make suitable changes in the connection of the armature turns, we double the E. M. F. of the armature (Par. 148). Thus, if we have an armature with a given number of turns on its surface and a given speed of rotation, in a bipolar field, and the E. M. F. obtained from the armature is 100 volts, then, if we change the field to a quadripolar frame, and suitably change the connection of the armature turns, the E. M. F. of the armature will be 200 volts. If, instead of changing the armature connections, we simply change the number of brushes from two to four, and suitably connect these brushes, we obtain only 100 volts as before, but as there are now four complete electric circuits through the armature, we have doubled the load which the armature can sustain without overheating, and, therefore, practically doubled the output of the armature, so that when we double the number of poles covering the armature, assuming the useful flux through each pole to be the same as before, we either double the torque-per-ampere directly, if the armature be series-connected, or we retain the torque-per-ampere with

FIG. 212.—QUADRIPOLAR CAR MOTOR WITH TWO FIELD COILS.

a multiple-connected armature and, by changing the winding, obtain a greater output from the motor.

351. There will, of course, be a limit to the number of poles which can be employed with any armature without increasing its diameter, since there will only be sufficient room for a certain number of poles carrying a given maximum flux, and also, since the difficulty of magnetizing a greater number of poles will be insuperable, either for want of space, or owing to increased magnetic leakage. The principle, however, is important.

352. The number of turns which can be utilized upon the surface of an armature is itself limited; first, by the resistance of the armature and consequent excessive heating under load; second, by excessive armature reaction and consequent sparking; and, third, in rarer cases, by the E. M. F. of the circuit,

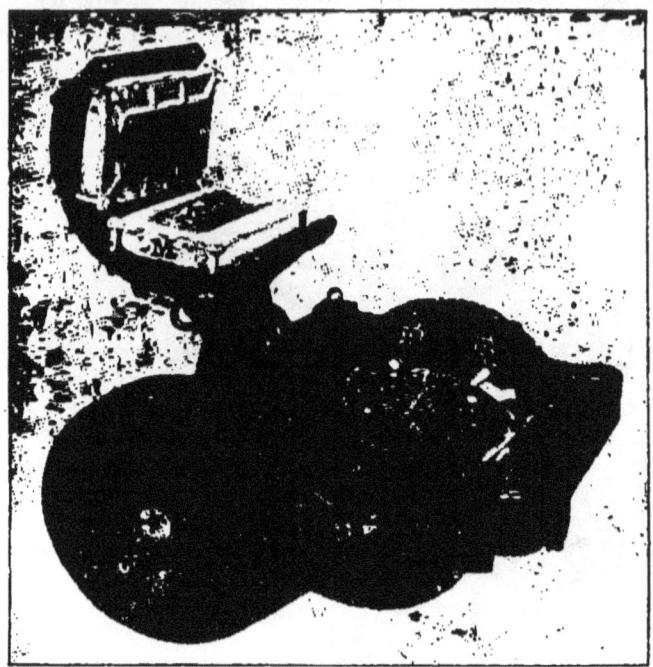

FIG. 213.—QUADRIPOLAR CAR MOTOR WITH FOUR FIELD COILS.

and, consequently, the unduly slow speed at which a powerful armature will run on such circuit.

353. The best embodiment of the foregoing principles in existing practice is found in a modern street-car motor. Here a powerful torque-per-ampere, with minimum weight, is desired in order to start a loaded car from rest up a steep gradient.

Two forms of such motors are shown in Figs. 212 and 213.

354. Fig. 212 shows a cast-steel quadripolar field frame with two magnetizing coils M, M. These produce not only poles at the opposite sides of the armature, in the cores over which they are wound, but also poles at the cylindrical projections *P*, *P*, which lie above and below the armature so that there are four complete magnetic circuits through the field frame and armature, two circuits through each magnetizing coil. The brushes *B*, *B*, are set 90 degrees apart on the commutator *C*. The armature *A*, is of the toothed-core type.

355. In Fig. 213 the same results are obtained with various detailed differences in mechanical construction. There are four poles around the armature, two of which, *P*, *P*, are seen in the raised cover, and two others are similarly contained in the lower half of the frame. Each of these poles is, in this case, surrounded by a magnetizing coil, *M*. *B*, *B*, are the brushes, set 90° apart from the commutator. The armature, *A*, is of the toothed-core type.

In both of these cases the magnetic circuits are as short as is practically possible, and the useful magnetic flux is as great as possible.

CHAPTER XXVII.

REGULATION OF MOTORS.

356. The requirements of a motor depend upon the nature and use of the apparatus which the motor is designed to drive. All these requirements, in relation to driving machinery, may be embraced under three heads; viz.,

(1.) Control of starting and stopping.

(2.) Control of speed, both as to constancy and as to variability.

(3.) Control of torque, both as to constancy and as to variability.

The above requirements are by no means met to an equal degree by the electric motor.

For example, the requirement of constant speed is much more readily dealt with than the requirement of variable speed.

357. The conditions under which motors have to operate may be divided into four classes; namely,

(1.) Constant torque and constant speed.

(2.) Variable torque and constant speed.

(3.) Constant torque and variable speed.

(4.) Variable torque and variable speed.

358. The first two conditions are readily secured, the third and fourth are only secured with difficulty. For example, a rotary pump belongs to the first class. Here the load is constant and the speed is presumably constant.

The second class comprises the greater number of machine tools, where the speed is constant but the activity is variable.

The third class embraces most elevators and hoisting machines.

The fourth class is well represented by street-car motors.

359. Any continuous-current electric motor will supply a constant torque at a constant speed when operated at a constant pressure. Thus, whether the motor be self-excited or separately-excited, and whether it be shunt-wound, series-wound or compound-wound, it will, if supplied with a constant pressure at its terminals, and assuming constant frictions in the machine, deliver a constant torque at a constant speed, and taking from the mains supplying it, a constant current strength, and, therefore, constant activity. The condition of constant torque and constant speed is one which is, therefore, readily dealt with by electric motors.

The above statement, however, is true only of single motors; for, if two motors, of any continuous-current type, be con-

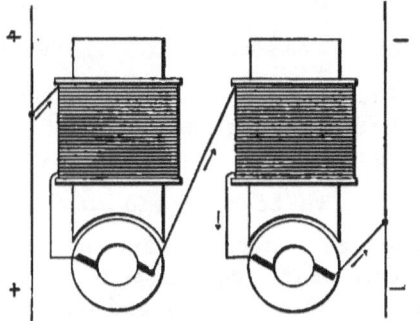

FIG. 214.—TWO SERIES-WOUND MOTORS COUPLED IN SERIES BETWEEN CONSTANT-POTENTIAL MAINS.

nected in series across a pair of constant-potential mains, they will be in unstable equilibrium as to speed under a given load. If the torque on each of the two machines in Fig. 214 were maintained absolutely equal; then, by symmetry, the two series motors represented would run at equal speeds, and absorb equal activities. But should the load on one accidentally increase, even to a small extent, above that of the other, the tendency would be to slow down the over-loaded motor and accelerate the other, so that it would be possible to have one motor at rest exerting a constant torque, and the other motor exerting the same torque at double its former speed. If, however, the two motors are rigidly coupled together to a countershaft, so that their speeds must be alike, then they will behave as a single motor. Consequently, a continuous-current motor employed for pumping or driving a fan, and which, therefore, has a constant torque to supply, will run at constant speed when supplied with constant pressure, whatever the type of motor may be.

360. The important requirement of constant speed under variable load is nearly met by a shunt-wound motor. It may be almost perfectly met by the compound-wound motor. It is not met, without the aid of special mechanism, by the series-wound motor.

361. Considering first the case of a shunt-wound motor, represented in Fig. 165, the speed at which the armature will run is $\dfrac{E}{e}$ revolutions-per-second (Par. 321), when at no load, provided that the friction of the machine is so small that we may safely neglect the drop of pressure in the armature running light. When the full-load current I amperes, passes through the armature, the speed will be reduced to $\dfrac{E - Ir}{e}$ revolutions-per-second, r, being the armature resistance in ohms.

Thus a particular shunt-wound, 110-volt motor has an armature resistance (hot) of 0.075 ohm, and its full-load output is 9 H. P. What will be its fall in speed between no load and full load, its no-load speed being 1,395 revolutions-per-minute or 23.25 revolutions-per-second?

Here, neglecting the armature torque and drop in pressure at no load, e, the dynamo power, or volts-per-revolution-per-second $= \dfrac{110}{23.25} = 4.73$. Its output at full load being $9 \times 746 = 6{,}714$ watts, and its armature efficiency, say, 0.84, the armature intake will be $\dfrac{6{,}714}{0.84} = 7{,}994$ watts $= 72.68$ amperes $\times$ 110 volts. The full-load armature drop will, therefore, be $72.68 \times 0.075 = 5.45$ volts, and the full-load speed $\dfrac{110 - 5.45}{4.73} = 22.1$ revolutions-per-second, approximately, or 1,326 revolutions-per-minute.

The drop in speed of this motor between no load and full load is, therefore, 69 revolutions-per-minute; or, approximately 5 per cent.

362. If the variation of speed due to the drop in the armature with the full-load current is greater than that which the conditions of driving will permit, then means may be adopted to reduce the value of e, at full load in the above formula, so as to increase the speed in compensation for the necessary drop. This is frequently accomplished by inserting resistance in the circuit of the field magnet so as to reduce its M. M. F., and, consequently, the useful flux which it sends through the armature. A rheostat in the shunt-field circuit, therefore, enables such regulation to be made by hand, as will maintain the speed of a shunt motor constant under all torques within its full load. For most commercial purposes the automatic regulation of the shunt motor is sufficiently close, the rheostat only being employed on special occasions. The larger the shunt motor the less the drop in speed which is brought about by the full-load current. Thus a 1-H. P. shunt motor will usually drop only 10 per cent. in speed at full load, a 10-H. P. motor 5 per cent., and a 100-H. P. motor, 3 per cent.

363. When a series motor is operated on a series circuit, as for example, on a series-arc circuit, some device is necessary which will regulate the speed of the motor. If no such device were provided, if the starting torque of the motor due to the constant current passing through it, exceeded the torque due to load and frictions combined, the motor would accelerate indefinitely in its endeavor to oppose by C. E. M. F. the passage of the current. If the load were of such a nature that the torque increased with the speed, as in the case of a fan, the speed might be automatically controlled, but, since, in driving machinery, the torque is nearly independent of the speed, a controlling mechanism becomes essential. One method by which this is accomplished is by rotating the rocker arm and brushes into such a position about the commutator, that the useful flux from the constantly excited series-wound field coils, passing through the armature coils, is virtually reduced by passing both into and out of the armature coils when the diameter of commutation is shifted, thereby neutralizing the electro-dynamic force on the windings. The method corresponds to that adopted for varying the E. M. F. of arc dynamos, in order to keep the current strength constant in the circuit, despite variations of load. (Par. 261.)

Fig. 215, represents a small series-wound $\frac{1}{6}$-H. P. motor for use on series-arc circuits and provided with a hand regulator to control the speed. The rocker arm, which supports the brush-holders, has a projection P, to which an insulating

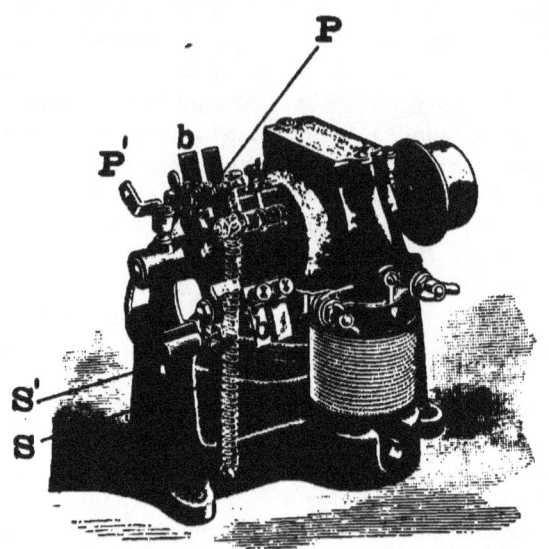

FIG. 215.—ONE-SIXTH H. P. MOTOR FOR ARC CIRCUITS WITH HAND OR TREADLE REGULATOR, ROTATING BRUSHES, AND AUTOMATIC CUT-OUT.

handle or treadle is attached. Under ordinary conditions, the spiral spring S, pulls the rocker arm, into the position shown, so that the brushes b, b, rest upon the commutator at a diameter at right angles to the diameter of neutral commutation in an ordinary bipolar motor, so that the torque of the motor will be reduced to zero. By rotating the rocker arm with handle or treadle against the tension of the spring S, so that the projection P, occupies the position P', the brushes are brought forward to the position b', of maximum torque, so that the speed of the motor may be controlled.

In the motor represented in Fig. 216, this rotation of the rocker arm is effected automatically by the aid of a centrifugal governor G, mounted at one end of the armature shaft.

When the motor is started, by throwing it into the series circuit by a switch, the brushes are at the diameter of neutral commutation or maximum torque. If the load torque is not too great for the armature to overcome, the motor will accelerate until the governor G, has lifted its wings to such a distance by centrifugal force against the tension of its

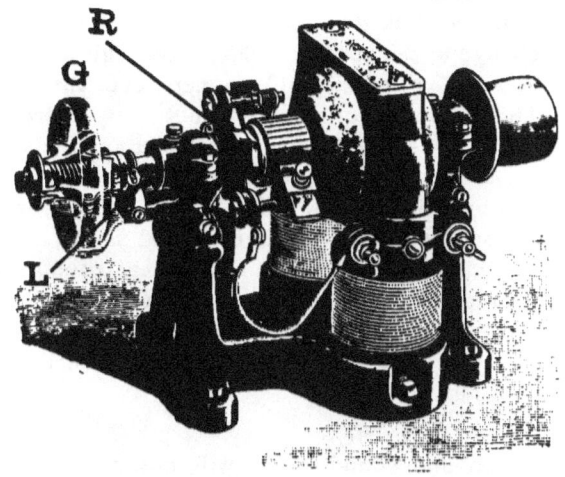

FIG. 216.—ONE-H. P. ARC MOTOR WITH AUTOMATIC GOVERNOR.

spring, that the lever L, following the motion of the governor, has pulled round the rocker arm and brushes to a diameter at which the torque of the armature is equal to that of the load.

364. In the ordinary motor the speed increases until the current strength I amperes passing the armature at the terminal pressure E volts, limits the intake, $E I$ watts, to the load activity and energy losses combined. In this motor the speed increases until the governor moves the brushes into such a position that the C. E. M. F., E volts, limits the activity of the constant current I amperes to the amount $E I$ watts, equal to the load activity and energy losses. The speed will, therefore, vary with the load by a small amount depending upon the sensibility of the governor.

Motors for series-arc circuits are not usually employed above 3 H. P. Owing to the high pressure which may exist upon their circuits, they may be dangerous to handle unless precautions are taken.

365. When a series-wound motor is employed across constant-potential mains, in the manner indicated in Fig. 164, the value of e, the dynamo power, or E. M. F. per-revolution-per-second, being equal to ϕw, varies with the torque or load, since any change in the current strength through the armature changes the M. M. F. of the field magnets, and, therefore, the flux ϕ. The tendency of a series motor is, therefore, to reduce its speed, as the torque imposed upon the motor is increased, and such a motor would run, theoretically, at an infinite speed on light load, if there were no frictions in the armature to be overcome. A shunt-wound motor, therefore, tends to drop in speed with load to an extent proportional to the drop of pressure in the armature. A series-wound motor falls off in speed with load, not only owing to the drop of pressure in the armature, but also owing to the increase in M. M. F. and flux.

366. A compound-wound motor will, however, maintain its speed practically constant under all loads, if the series winding on the field coils be so adjusted that the increase in current strength through these coils and the armature shall diminish the M. M. F. of the field magnets to the degree necessary to compensate for the drop of pressure in the armature winding. The connections of such a compound-wound motor are the same as for the compound-wound dynamo shown in Fig. 166.

367. Although a series-wound motor is unfitted for maintaining a constant speed on constant-potential mains with variable torque, yet it is possible to connect two series-wound machines of the same type and character together, one acting as a generator and the other as a motor, and to obtain a nearly constant speed of the motor by compensatory changes in the E. M. F. of the generator automatically brought about by the variations of load. This case, however, can only apply to a single motor driven by a single generator, and is, therefore, inapplicable to a system of motors driven by a single generating source.

368. Figs. 217 and 218 are diagrams taken from actual tests of two small 500-volt, ½-H. P. motors, of good construction and well-known manufacture, one being a series-wound motor and

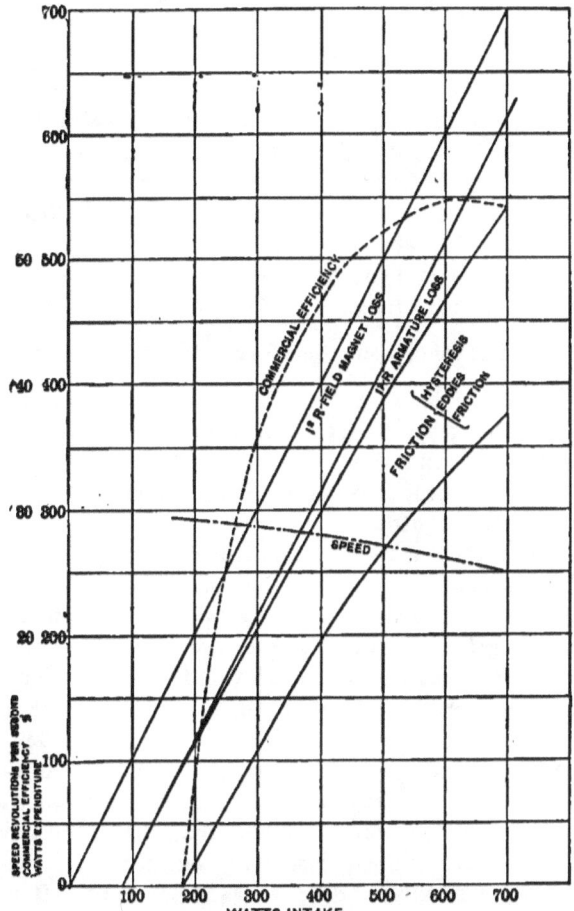

FIG. 217.—TEST DIAGRAM OF A SHUNT-WOUND ONE-HALF H. P. MOTOR SHOWING DISTRIBUTION OF ACTIVITY.

the other a shunt-wound motor. The armatures of the two machines and also their field frames were practically identical, the only essential difference between the two being in the field winding. The weight of the machines was 105 lbs. each, that of the armature nearly 22 lbs. The resistance of the armatures was 40 ohms each, and the resistance of the fields 3,680 ohms for the shunt-wound, and 37.5 ohms for the series-wound, machine.

In these diagrams, the ordinates represent the expenditure of activity in the field windings, armature windings, frictions (including hysteresis, eddy currents, and mechanical frictions), and output at the shaft. The abscissas represent the intake in watts. Thus, referring to Fig. 217 for the shunt-wound machine, it will be seen that when delivering full load, or 373 watts, the machine absorbed 690 watts, expending 90 in the field magnets, as $I^2 R$, 67 watts in the armature as $I^2 R$, and 160 watts in total frictions. The commercial efficiency of the machine at full load, was, therefore, $\frac{373}{690}$ or 54 per cent. The speed of the machine falls from 29.2 to 25 revolutions-per-second, or from 1,752 to 1,500 revolutions per minute, a drop of 14.4 per cent., and this drop is closely proportional to the output. The highest commercial efficiency reached was 55 per cent. at 340 watts output.

Taking now the series-wound machine referred to in Fig. 218, it will be observed that the field loss is much smaller, particularly at light loads, owing to the fact that it increases with the current strength, and practically disappears when the current strength is very small. Owing to this fact it will be observed that the commercial efficiency of this machine is greater throughout than that of the shunt machine. At a delivery of 340 watts, the intake was 600 watts, expended as follows: 57 watts in the magnets, 63 in the armature, and 140 watts in frictions. It will be seen, however, that the speed falls from 38.5 to 21.5 revolutions-per-second, or from 2,310 to 1,290 revolutions-per-minute, a drop of 44.2 per cent. It is clear, therefore, that a series-wound machine is, in small sizes, cheaper to construct than a shunt-wound machine, since it employs only a few turns of coarse wire instead of many turns of fine wire in its field coils. It also has a slightly higher efficiency. It also dispenses with the use of a starting rheostat in the armature, but has the disadvantage of possessing a much greater variation in speed under variations of load.

369. As already mentioned, the condition of constant torque and variable speed is one which it is much more difficult for the electric motor to meet. If it were possible to vary the

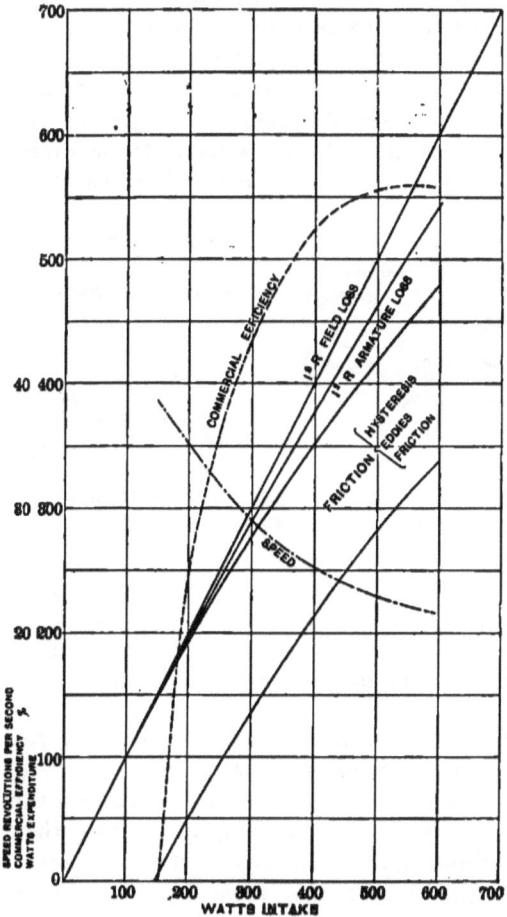

FIG. 218.—TEST DIAGRAM OF A SERIES-WOUND ONE-HALF H. P. MOTOR SHOWING DISTRIBUTION OF ACTIVITY.

useful magnetic flux through the armature within wide limits, the method of varying the M. M. F. of the field magnets would effect the result desired. While, however, it is possible to produce a variation of speed in the ratio of 3 to 1, by varying the M. M. F.; that is to say, while motors have been constructed, under special conditions, which will run, say at from a maximum of 900, to a minimum of 300 revolutions-per-minute, merely owing to variation in the M. M. F. of their fields, yet such a range is only obtained with great difficulty, owing to the fact that magnetic saturation is reached at maximum M. M. Fs. in the iron constituting the magnetic circuit, and that when the field flux is greatly reduced, the armature reaction at full load is liable to be excessive, with heavy sparking at the commutator. The maximum range of

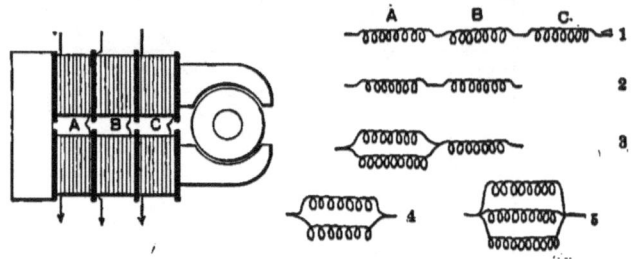

FIG. 219.—DIAGRAM SHOWING ONE METHOD OF SERIES-PARALLEL FIELD EXCITATION IN A STREET-CAR MOTOR.

speed in an ordinary shunt motor, brought about by field regulation, is only about 25 per cent., so that a motor whose maximum safe speed is 1,000 revolutions-per-minute, can be reduced to minimum of about 750 revolutions.

370. The M. M. F. of a motor field may be varied electrically in two ways; namely, by altering the current strength through the field coils as a whole, by inserting a varied resistance in their circuit; and second, by altering the action of certain portions of the field coils relatively to other portions, as, for example, by changing them from series to parallel, or the reverse. In shunt-wound motors, the regulation is usually effected by the introduction of a field rheostat. In series-wound motors it is usually effected by varying the number or arrangement of the field coils. Thus the arrangement for connecting the field coils of a particular form of street-car motor is represented in Fig. 219. It will be seen that there are three coils on each limb of the field, but each pair is permanently connected as shown, so that electrically there are only three coils, *A*, *B* and *C*. By the action of the controlling switch, these coils may be connected as shown in the diagram.

In Position 1, all three coils are in series, making the relative M. M. F. 3 and the relative resistance 3.

In Position 2, one coil is short circuited, making the relative M. M. F. 2 and the relative resistance 2.

In Position 3, two coils are connected in parallel, making the relative M. M. F. 2 and the relative resistance 1.5.

In Position 4, two coils only are connected in parallel, making the relative M. M. F. 1 and the relative resistance 0.5.

In Position 5, all three coils are connected in parallel, making the relative M. M. F. 1 and the resistance 0.333.

Fig. 220 represents the characteristic curve of a particular motor of this character, with the flux in megawebers, passing

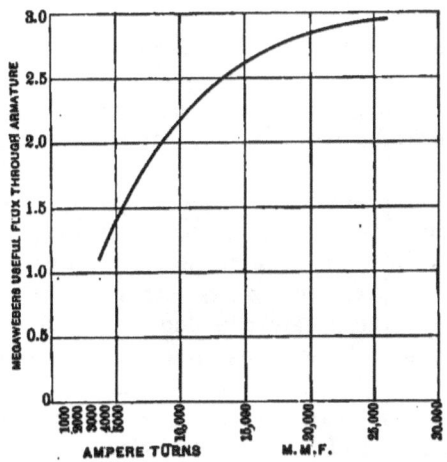

FIG. 220.—CURVE OF MAGNETIC FLUX THROUGH ARMATURE IN RELATION TO M. M. F. OF FIELD MAGNETS.

through the armature with different excitations of the field magnets, expressed in ampere-turns. With the aid of this curve it is possible to estimate the range of speed which can be obtained by connecting the coils in different arrangements. For example, at half load of 7½ H. P., or say 5,600 watts output, and an efficiency of say 0.8, the activity absorbed would be 7,000 watts, or 14 amperes at 500 volts pressure. There are, approximately, 2,100 turns in the field coils, or 700 to each pair, so that with all in series, the total M. M. F. would be $14 \times 2,100 = 29,400$, which might produce a flux of 2.9 megawebers through the armature. With all the coils in parallel, the M. M. F. would be three times less or 9,800, and the flux 2.12 megawebers. The ratio of speed, therefore, would be $\frac{2.9}{2.12} = 1.368$, so far as regards the effect of change in magnetic flux through the armature. In practice, the speed would vary in a somewhat greater ratio, owing to the influence of greater drop in the field magnets when connected in series than when connected in parallel. We may consider, therefore, that at light loads the influence on the speed of varying the field coil connections is considerable, but at heavy loads the influence is relatively small.

371. We have seen how the speed of a motor can be controlled within certain limits by varying the magnetic flux usefully passing through its armature. The same results can be effected by introducing resistance into the armature circuit.

372. If the constant torque imposed upon the motor is such as requires a current of I amperes to pass through its armature, while a given constant magnetic flux is produced by the field, and if E, be the pressure in volts across the main leads, and r, the resistance of the armature in ohms, the drop in the armature will be Ir volts, and the armature of the motor must develop that speed which will produce a C. E. M. F. of $(E - Ir)$ volts. If it be required to reduce this speed to say, one half, then the total resistance of the armature circuit must be increased to R ohms, in such a manner that $E - IR = \frac{E - Ir}{2}$, so that $R = \frac{E + Ir}{2I}$. While this plan is theoretically effective, it is practically objectionable, because, in the first place, it wastes energy by the introduction of the additional resistance $(R - r)$ ohms, the amount of activity wastefully expended in such resistance being $I^2 (R - r)$ watts. In the second place, a comparatively small accidental variation in the torque, which we have hitherto supposed constant, would effect a large variation in the speed, owing to the varying drop in the added resistance. Again, a powerful motor requires a powerful current strength to be supplied to it, and a large expenditure of energy is necessary in order to greatly reduce its speed in this

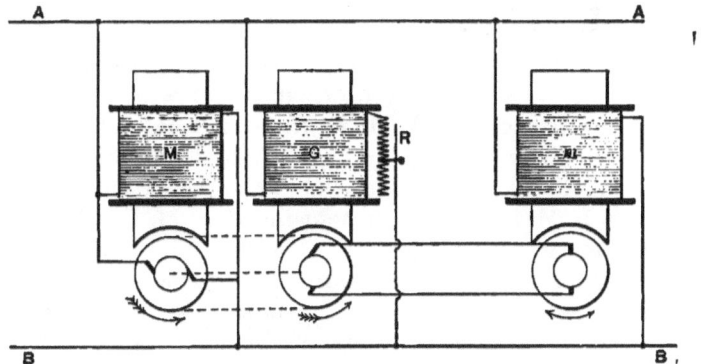

FIG. 221.—SYSTEM OF PRIME MOTOR, GENERATOR, AND WORKING MOTOR FOR CONTROLLING THE SPEED AND DIRECTION OF THE WORKING MOTOR UNDER CONSTANT TORQUE.

manner, requiring the use of bulky and expensive resistances, to dissipate the heat developed. For these reasons this method of maintaining the speed constant is seldom employed.

373. It has been found so difficult in practice to vary the speed of a motor at constant torque between full speed and rest, without loss of efficiency, that in cases where complete control is imperative, as in some rolling mills, where the machinery has to run occasionally at a definite very low speed, and at other times at full speed, a method, which is represented in Fig. 221, has been invented and applied. Here M, is a shunt-wound motor, connected across a pair of supply mains, $A A$, $B B$, and, therefore, running at practically constant speed under all conditions of use. The armature of this motor is connected directly, either by a belt or by a rigid coupling, to the armature of the generator G, whose field magnets are excited through a rheostat R. The generator armature consequently runs at a practically constant speed under all conditions of service. The E. M. F., which this generator armature develops, depends, however, upon the excitation of its field magnets, which is regulated by the rheostat R, so that, when no current passes through the generator field coils, the E. M. F. of its armature is nearly zero, while, when full current strength passes through the field coils, the E. M. F. of the generator is at its maximum. The brushes of the generator are directly connected with the brushes of the working motor m, whose field magnet is constantly excited, and the speed of the armature m, will be controlled directly by the E. M. F. of the generator G. If the generator is fully excited, the E. M. F. at the terminals of the motor m, will be a maximum, and the speed of the motor to meet this E. M. F. with a corresponding C. E. M. F. will also be a maximum, while if the generator has its excitation removed, the armature of the motor m may come almost or quite to a standstill. If necessary, the connecting wires between the armatures of G and m, can then be reversed so that the direction of m's rotation can be reversed.

374. The fact that this combination of machines operates satisfactorily without excessive sparking at the commutator of the generator, often occasions some surprise to those who are accustomed to varying the field excitation of generators and motors, under ordinary conditions, since it is known that, in general, when a generator, and particularly a motor, has its field magnets considerably weakened, a violent sparking is apt to be produced at the commutator. It is to be remembered, however, in this case, that the armature of the weakened generator G, is never permitted to send more than the full-load current strength, which is required to overcome the full-load torque, while on the contrary, if this machine were employed across constant-potential mains as a motor and the magnetic flux through the armature was considerably weakened, the current strength which would pass through the armature would be, probably, much in excess of the full-load current, with a corresponding tendency to produce excessive armature reaction and sparking.

375. Although the preceding combination of apparatus effects the desired result of varying or reversing the speed of the motor at will, under constant or even under variable torque, within the limits of full load, yet it has the double disadvantage of requiring the installation of three times the amount of machinery which would otherwise be necessary, and of having a considerably reduced efficiency of operation. If, for example, the motor M, has to be a 10-KW machine, then the generator G, must at least have a capacity of 10 KW, and at least an equal capacity will have to be given to the prime motor M; so that 30 KW of machinery are installed where but 10 are directly brought into use. Again, if the commercial efficiency of each machine were 83 per cent. at full load, the commercial efficiency of the combination, under full load, would be, approximately, $0.83 \times 0.83 \times 0.83 = 0.572$, so that the combination would have a full-load efficiency of 57.2 per cent. At light loads the combination efficiency would be still lower; for example, if at half load the efficiency of each machine were 75 per cent., the combination efficiency would be 42.2 per cent. On the other hand, however, the introduction of resistance into the armature circuit of a motor, in order to reduce its speed, would probably effect as low or even a lower efficiency. It is evident, therefore, that in this direction the electric motor shows its weakest side.

376. The fourth condition of working; namely, under variable torque and variable speed, differs from the last only in the variability of the torque. This being, as we have seen, the condition of working with street-car motors, it is probably one of the most important conditions to be met. It is met within the limits of practical requirements in street-car motors, partly by controlling the field magnets, and partly by the introduction of resistance into the armature circuits. This resistance may be added either through the series windings of the field coils, or by the direct insertion of external resistance. The problem, however, of controlling within full range the speed of a single continuous-current motor, under varying torque, with high efficiency, is, strictly speaking, yet unsolved.

377. In some cases two motors are rigidly coupled together so that they may have their armatures connected in series or in parallel. In the first case they divide the pressure of the

C. E. M. F. between them, so that their speed will be a minimum under that condition. In the second case they each take the full pressure, and so yield the maximum speed. At slow speed, however, when connected in series, it is evident that the activity of the combination will be EI watts, since each machine can now take I amperes, E, being the pressure between the mains, in volts. At full speed, since each armature can take I amperes, the available activity will be $2EI$ watts. The combined torque, for the full-load current through each armature, will be the same whether they are in parallel or in series.

CHAPTER XXVIII.

STARTING AND REVERSING OF MOTORS.

378. If a series motor be at rest, and be connected directly across the mains, then if the resistance of the armature and magnet coils together be R ohms, the current strength passing through the motor tends to become $\frac{E}{R}$ amperes, E, being the E. M. F. in volts at the supply mains. Thus, if a 1-H. P. series-wound motor has a resistance in the armature of 0.5 ohm, and a resistance in the field coil of 0.5 ohm, the total resistance in the machine will be 1 ohm, so that the first tendency is to produce a current strength of $\frac{110}{1} = 110$ amperes, as soon as the machine is connected with the circuit, assuming the mains to have a constant pressure of 110 volts, whereas the full-load current strength of the machine will be about 10 amperes. As soon as the armature has become able to develop its full speed, the motor will generate such a C. E. M. F. as will limit the current through it to that required to expend the energy it wastes and delivers. The rapidity with which the armature will reach its full speed depends upon the load connected with it, upon the inertia of the armature and of its load, as well as upon the current strength entering the armature. Moreover, owing to the self induction, or inductance, of the field-magnet coils, it is impossible to develop the full current strength immediately in them, even assuming that the armature were to remain at rest. As soon as the current excites the field magnets, the flux they produce, passing through the magnetic circuit, develops in the field coils a temporary C. E. M. F., which has a powerful influence in checking the first inrush of current into the armature during the first half second or second of time. For this reason, a series-wound machine is much more safely started from rest to full speed than a shunt-wound machine, in which the armature has to be connected directly across the mains.

379. In all except the smallest machines of the shunt-wound type, it is necessary to insert some resistance in the armature circuit when starting from a state of rest, so that the drop produced in such resistance by the starting current may limit the amount of current passing through the armature. For this purpose special rheostats, called *starting rheostats*, are inserted in the armature circuit. Since they are only intended to carry the current during the time that the motor is coming up to speed, they are not usually designed to carry the full current strength of the motor indefinitely, and, therefore, a starting rheostat should never be maintained constantly in circuit. Fig. 222 represents a form of starting rheostat employed with shunt-wound motors. Here a number of coils or spirals of

FIG. 222.—STARTING RHEOSTAT.

galvanized iron wire, are mounted in a fire-proof frame under a cover of slate or composition, on which a number of contacts are arranged in a circle. Fig. 223 represents the manner in which such a rheostat is connected in the armature circuit.

380. If it becomes necessary, as we have shown, to insert resistance into the circuit of a shunt-wound motor armature, in order to start it from rest, it is still more necessary to insert resistance into the armature circuit, in order suddenly to reverse its direction of motion. When the armature terminals of a shunt-wound motor are suddenly reversed, relatively to the mains, while the field magnet coils remain permanently excited, the E. M. F. of the armature due to its speed, which was, before the reversal, a C. E. M. F., tending to check the passage of current strength through its windings, becomes now a driving E. M. F., tending to increase the current strength passing through it from the mains. The effect of a sudden reversal in a shunt-wound motor armature is, therefore, practically equivalent to suddenly throwing the armature across a pair of mains having double the pressure of those actually employed, and

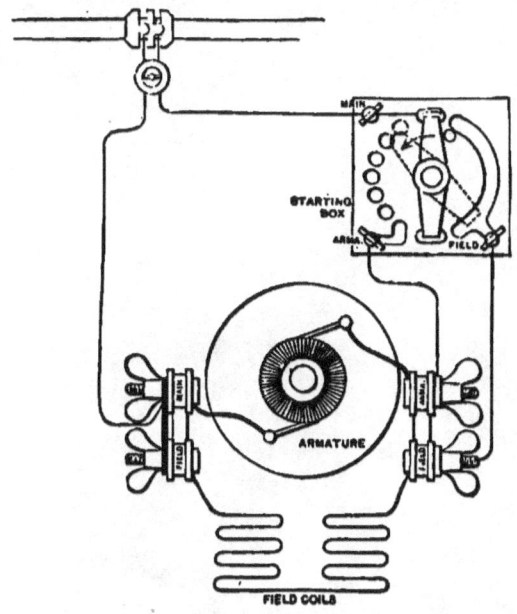

FIG. 223.—CONNECTIONS OF STARTING RHEOSTAT WITH SHUNT MOTOR.

with the attending consequences of an enormous overload of current strength, which first checks, and then reverses, the direction of armature rotation.

381. Various devices are employed for preventing a motor armature from being injured by the sudden reversal of its terminals with the mains. At the time when armatures were almost all of the smooth-core type, damage was frequently done by shearing the wires off the armature core under the very heavy

FIG. 224.—FORM OF AUTOMATIC SWITCH.

electro-magnetic stresses thus brought to bear upon them during rotation. When toothed-core armatures became generally used this danger practically disappeared, but the danger of damaging either the insulation of the wires, or the mechanical framework of the armature, or of burning out some of the con-

STARTING AND REVERSING OF MOTORS. 301

ductors, still remains. A *starting coil* is frequently employed with street-car motors which consists of a coil of strip-iron conductor, having a hollow interior, so that it contains a large

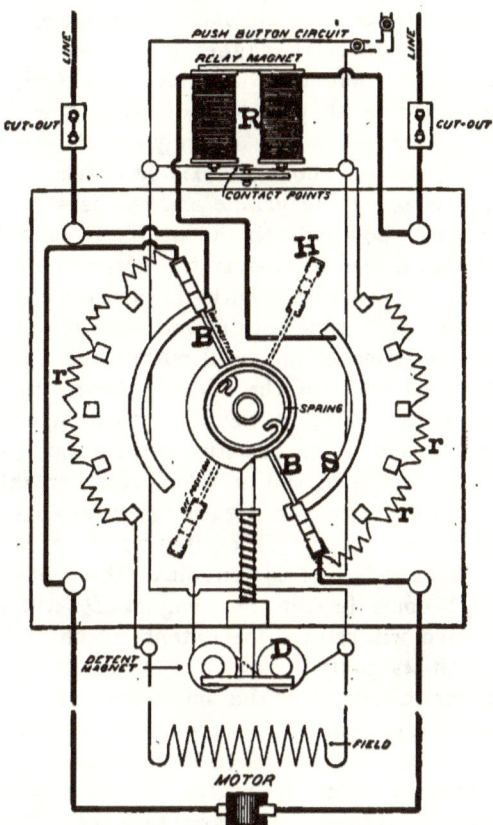

FIG. 225.—CONNECTIONS FOR AUTOMATIC SAFETY SWITCH AND STARTING RHEOSTAT.

magnetic flux when excited. The C. E. M. F. suddenly developed from such a coil, on being magnetized, is sufficiently great, to check, for the moment, the first rush of current, and such a coil may be called an *inductance coil*.

382. Fig. 224, represents the form, and Fig. 225, the diagrammatic connections of a particular automatic switch and starting rheostat sometimes employed with large motors. The larger the motor the more expensive does any accident become which may happen to its armature, and the more economical it becomes to take precautions against such accidents. Referring to the figures, it will be seen that the mains or line wires are connected directly to two circular contact segments S, S, through the coils of a relay magnet R. When the handle H, is in such a position that the two contact bars B, B, rest in the intermediate position, they lie out of contact with the segments, and the current is then entirely cut off the motor. A powerful spring, wound about the axis on which the handle H, moves, tends to bring the handle and the bars B, B, back to this zero or "off" position. If the handle is pressed forward in the clockwise direction against the pressure of its spring, the line wires are connected with the armature through the resistance coils r, r, r, which are wound upon spools of insulating and non-inflammable material within the box, and also through the field coils of the motor. When the handle is pushed completely around to the "on" position, the extra resistances are cut out of the armature circuit and the armature thus becomes enabled to run at full speed. In this position the handle is prevented from returning to zero and is kept in place by the detent magnet D, excited by the current passing through the field coils. If the circuit of the field coils should accidentally become broken, the magnet D, will release its armature, which will release the detent, which will allow the handle H, with its contact bars B, B, to return to the "off" position, under the action of the spiral spring; or, should the armature current become excessively strong, thereby endangering the armature, the relay magnet will attract its armature, which will thereby short-circuit the detent magnet, and the same result will follow. The armature will, therefore, be stopped by any overload, and will be cut out of circuit by any accidental cessation of the current in the field. By means of a push-button circuit, the armature can be brought to rest, by pressing a push button placed at any distance from the machine.

383. All the phenomena of armature reaction which we have traced in connection with dynamos in Pars. 198 to 223 are presented by motors, with the exception that the direction of the M. M. F. of the armature, relatively to the field magnets, is reversed; that is to say, a motor runs so that the magnetic flux produced by its armature tends to pass through the pole which the armature approaches; *i. e.*, the *leading pole*, instead of the *trailing pole*, or that from which it is forced in the dynamo. With this exception all the effects of sparking and cross-magnetization present themselves equally in motors as in dynamos. The diameter of commutation in a generator has to be advanced in order to obtain a sparkless position; in other words, a *lead* has to be given to the brushes, while in a motor the diameter of commutation has to be retrograded to arrive at the same result; in other words, a *lag* has to be given to the brushes.

384. In order to reverse the direction of rotation of a motor, a single rule has to be borne in mind; namely, the M. M. F. either of the field or of the armature must be reversed. If the M. M. F. of both field and armature be simultaneously reversed, the direction of rotation of the motors remains unaltered.

385. Fig. 226 is a complete diagram showing the relations which exist between the direction of rotation and the direction of current in the field and armature of different machines. The horizontal row on the top represents separately-excited machines; the next lower row, shunt-wound machines, and the lowest horizontal row, series-wound machines. The first vertical column, No. I, on the right, represents generators. Column II, next in order to the left, represents the action of these machines as motors, when mounted in connection with the mains, but not supplied with sufficient driving power to maintain the machines as generators. Column III represents the effect of reversing the connection of the armature when the machine is acting as a motor. Column IV represents the effect of reversing the field connections instead of the connections of the armature. Column V represents the effect of reversing both field and armature connections, which is equivalent to reversing the entire machine relatively to the mains. The large arrow on the field coil represents the direction of

304   *ELECTRO-DYNAMIC MACHINERY.*

the M. M. F., or of flux through the field. The large arrow on the armature represents the direction of the M. M. F. in the armature, due to the current. The small arrow in the centre of the armature represents the direction of the arma-

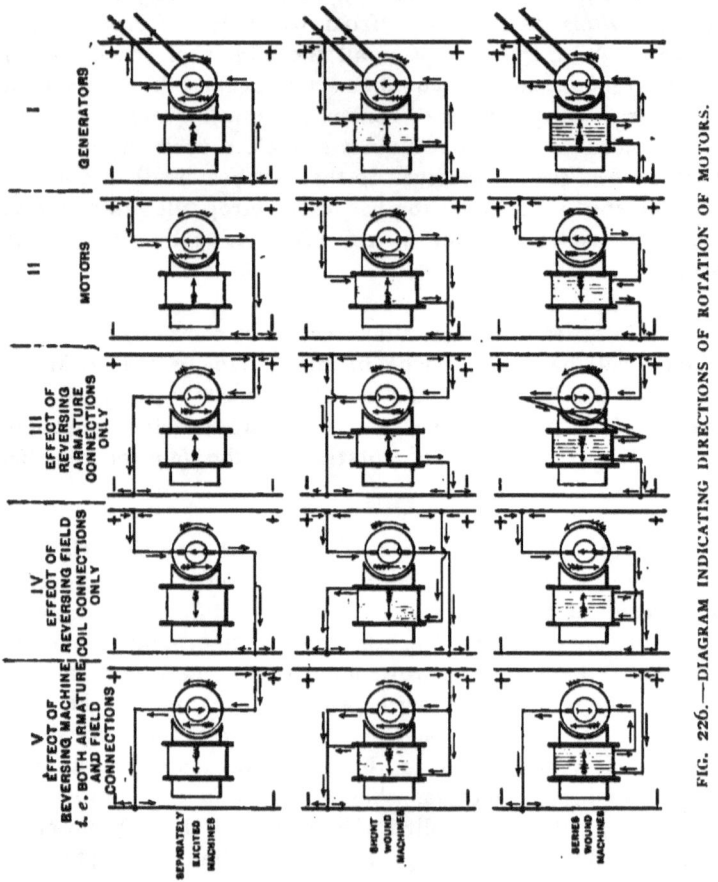

FIG. 226.—DIAGRAM INDICATING DIRECTIONS OF ROTATION OF MOTORS.

ture E. M. F., relatively to the circuit, and the curved arrow, outside the armature, represents the direction of rotation of the armature.

386. Referring to the line or row of separately-excited machines, in Column I, each machine appears as a generator, rotated by the driving belt in the direction of the curved arrow. The E. M. F. of the armature is in the direction of the current through the armature, and the mains are supplied with current from the brushes, as shown. If the driving belt be suddenly thrown off the armature pulley, the machine will run for a few moments by its inertia, still supplying current to the mains, until the power so expended has absorbed the surplus energy of motion of the armature, when the speed and E. M. F. of the armature will diminish, until the E. M. F. is exactly equal to that between the mains, which are assumed to be maintained at a constant difference of potential by another source of supply. At this moment there will be no current through the armature. If there were no friction in the armature, this condition might be retained indefinitely, but since every machine must expend energy against frictions, the speed of the armature continues to slacken, and the E. M. F. in the armature falls below that in the mains. Current will then pass back from the mains through the armature, as shown in Column II, reversing the M. M. F. of the armature, but maintaining the same direction of rotation. The machine is now rotated as a motor, absorbing energy from the mains, and the E. M. F. of the armature is now a C. E. M. F., as shown by the opposition between the directions of the small arrow in the centre of the armature, and the arrows representing the direction of current through the armature. Consequently, a separately-excited machine runs in the same direction as generator or motor, if no change is made in the armature or field connections. If the armature connections be reversed, as represented in Column III, or if the field connections be reversed, as represented in Column IV, the direction of rotation of the armature is reversed; but, if both field and armature connections be reversed, as in Column V, the original direction of rotation is retained.

387. 'In the shunt-wound machines, represented in the second row, practically the same conditions are observed to follow; namely, if the driving belt be thrown off the pulley of the machine acting as a generator, when connected to constant-potential mains, current will pass through the armature in the opposite direction to that which passes when the machine is a generator, thus reversing the M. M. F. of the armature, but maintaining the direction of rotation. Reversing either the field or the armature, reverses the direction of rotation, but reversing the entire machine; *i. e.*, both field and armature, has no effect upon the direction of rotation.

388. The third row; viz., that of series-wound machines, differs, however, essentially from the foregoing. Here, it will be observed, that if the belt be thrown off the generator, as soon as the E. M. F. of the armature is brought down to that existing between the mains, no current passes through the mains and the field magnets lose their excitation. It will follow from this that the E. M. F. of the armature will very rapidly disappear, and a large rush of current will pass through the armature from the mains, reversing the direction, not only of the armature M. M. F., but also of the field M. M. F., so that the machine is first brought to a standstill, and then rotated in the opposite direction. It is clear, therefore, from this consideration, why series-wound machines are never employed as independent units, in parallel, for supplying a system of mains; for, if by any acccident the engine driving a series-wound generator failed to maintain the E. M. F. of its armature above that of the mains, the machine would become a short circuit upon the mains, and an enormous rush of current, with a correspondingly violent mechanical effort, would be brought to bear upon the machine, tending to reverse its motion and drive the engine backward.

389. If the series-wound machine be considered as running in the direction represented in Column II, and the armature connections are then reversed, or the field magnet connections reversed, as in Columns III and IV, the direction of rotation of the armature will be reversed, or restored to the direction of rotation as a generator ; while, if both field and armature be reversed, as shown in Column V, the direction of rotation will be the same as in Column II.

390. It is evident, therefore, from an inspection of the diagram, that it is only necessary either to reverse the direction of the M. M. F. in the armature or in the field, to reverse the direction of rotation of the motor, and that the relative direction of the M. M. F. in field and armature is opposite in a motor to what it is in the same machine as a generator. For this reason the leading pole-pieces of a machine, when operating as a generator, and the following pole-pieces when operating as a motor are weakened by armature reaction.

391. In practice, it is always the connections of the armature of a machine which are reversed, in order suddenly to reverse the direction of its rotation, for the reason that the inductance of the armature being usually much less than that of the field, the change is more readily effected, and with less danger of injuring the machine by an excessive rise of pressure. On the other hand, if the machine be brought to rest and disconnected from the circuit, it may be just as convenient to reverse the field magnet connections as the armature connections, in order to effect a reversal of rotary direction when the machine is next started.

392. In all cases it has to be remembered that it is dangerous to break the circuit of the field magnets of a motor when in operation, not only because by so doing the M. M. F. of the field is almost entirely removed, and thereby the armature is unable to develop a C. E. M. F., becoming practically a short circuit on the mains; but also, because the powerful E. M. F. generated in the field coils by self-induction, when their circuit is interrupted, may find a discharge through the armature insulation, in such a manner as to pierce the same and permanently injure the armature. The same remarks apply to the operation of machines as generators. The field magnet connections should always be the first to be completed, and the last to be interrupted, when the machine is operated in either capacity.

393. In some cases, it is possible for the M. M. F. of the armature to overcome that of the field magnets, and actually to reverse the direction of magnetic flux through the magnetic circuit of the machine. For example, if a shunt-wound machine be operating alone, and supplying a system of mains, it is possible for a very powerful current passing through the armature to produce such an armature reaction as shall effect a large C. M. M. F. in the magnetic circuit of the machine, and so reverse the magnetic flux in the circuit. As soon as this is effected, the E. M. F. of the armature will be extinguished and the machine will cease to send a current. This effect is distinct from the tendency of shunt-wound generators to lower their E. M. F. under heavy loads, by reason of the drop in the armature, and its effect upon the excitation of the field magnets. It can only happen when the brushes of the machine are given a considerable lead; for, if the brushes be maintained at the neutral point midway between the poles, it will be impossible for the armature reaction to produce a dangerously large C. M. M. F. in the main magnetic circuit. Such accidents have, however, taken place in central stations with types of generator in which the armature reaction and lead of the brushes at full load is considerable. For this reason it is preferable to excite the field magnets of large central station generators from independent machines, when possible.

394. In motors, which are required to have their direction reversed, it is necessary that the brushes shall rest upon the commutator in such a position as shall permit of this reversal of direction without danger. Carbon brushes are employed with practically all 500-volt generators and motors, and with such machines for lower pressures as will permit of the passage of the full-load current through the carbon brushes without dangerously overheating them. Their advantage is that they wear evenly, lubricate the surface of the commutator, and are readily replaced. Their only disadvantage is their high resistivity, and the noise they are apt to make if the commutator surface is not perfectly uniform.

CHAPTER XXIX.

METER-MOTORS.

395. It sometimes becomes necessary to design a motor, whose speed shall be proportional to the current strength passing through it. This problem arises in devising *motor-meters* for determining the quantity of electricity supplied to a customer from a pair of constant-potential mains, as in electric lighting. The motors employed for this purpose are of very small sizes. We propose to consider the conditions under which the speed of the motor shall be proportional to the driving current strength.

396. Fig. 227 represents a pair of constant-potential mains, marked $+$ and $-$, with a small motor M, designed to measure the current strength supplied to the incandescent lamps, $L\ L$, with which it is connected in series. It is evident that the current which passes through the motor armature will vary directly with the number of lamps which are turned on. The connections of the motor field magnets are not shown. These magnets may be constantly excited from the mains, thus virtually constituting a separately-excited field; or, a permanent magnet field may be employed for this purpose. In either case the strength of the field flux may be considered as independent of the load.

397. We know that (Par. 313) if i, be the current strength passing through the armature in amperes, Φ, the field flux, in webers, usefully passing through the armature, and w, the number of turns on the armature, counted once completely around; the torque-per-ampere, which will be exerted about the armature shaft will be

$$\tau = \frac{\Phi w}{10.2\pi} \quad \text{cm.-dynes per ampere.}$$

If no load except friction were imposed upon the armature, that is to say, if it were free to run without retarding torque, beyond a frictional torque of f cm.-dynes, due to mechanical and electric frictions, then the speed which the motor would attain, as soon as the first lamp was turned on, would be very great, assuming that the torque $i\tau$, was sufficient to start the motor; for, the friction f, would be practically constant at all speeds, and if $i\tau$, be greater than f, the accelerating force being greater than the retarding forces, will continually increase the speed of the motor until the C. E. M. F. of the armature reduces the current strength to that which is needed to exactly neutralize the retarding torque. Such a small motor, therefore, if unloaded, would tend to run at a very

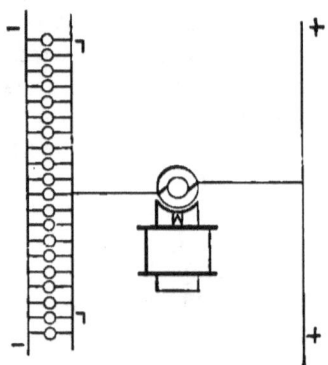

FIG 227.—MOTOR ARMATURE IN CIRCUIT WITH INCANDESCENT LAMPS.

high speed and to reduce the pressure at the terminals of the lamp.

398. It is also evident that the resistance of the armature must be sufficiently small, in order that the drop and C. E. M. F. in the armature, produced by the full-load current, shall not be greater than say one per cent. of the total pressure at the mains. Let us assume that we are able to impose a load or torque upon the motor proportional to its speed. If n, be the number of revolutions-per-second made by the motor, τ, the load torque in cm.-dynes will then be $\tau = a\,n$, where a, is a constant quantity. Under these conditions, the speed which the motor will attain will be determined by the equality of the driving and resisting torques or $i\tau = a\,n + f$. From which

$$n = \frac{i\tau - f}{a} \text{ revolutions per second} = \frac{\tau}{a} - \frac{f}{a}.$$

399. For example, suppose a small motor to be connected as shown in Fig. 227, in circuit with 20 incandescent lamps, each taking one half ampere from a pair of mains supplied with 110 volts pressure. The full-load current will be 10 amperes, and, if the resistance of the armature be 0.1 ohm, the drop of pressure in the armature at full load will be 1 volt. If the torque τ, of the motor be 200 centimetre-grammes, or, approximately, 200,000 centimetre-dynes per ampere of current, also if the torque due to frictions be 75 centimetre-grammes, or, approximately, 75,000 centimetre-dynes, and the torque due to load be 120 centimetre-grammes, or, approximately, 120,000 centimetre-dynes-per-revolution-per-second, then, if one lamp were turned on, the current through the armature would be 0.5 ampere. The starting torque would be 100 centimetre-grammes, the resisting torque of friction, 75 centimetre-grammes, and the motor would therefore start under a resultant torque of 25 centimetre-grammes. It would accelerate until a speed of 0.208 revolution-per-second was attained, when the resisting load torque would be 0.208 × 120 = 25 centimetre-grammes. Proceeding in this way, we can determine what the speed of the motor would be with any current strength as follows:

Lamps.	Current amperes.	Moving-Torque cm.-grammes.	Resisting Friction cm.-gms.	Torque Speed cm.-gms.	Speed of motor rv. per s.	Speed per lamp rv. per s.
1	0.5	100	75	25	0.21	0.21
2	1.0	200	75	125	1.04	0.52
4	2.0	400	75	325	2.71	0.677
6	3.0	600	75	525	4.375	0.729
8	4.0	800	75	725	6.04	0.755
10	5.0	1,000	75	925	7.71	0.771
12	6.0	1,200	75	1,125	9.375	0.781
14	7.0	1,400	75	1,325	11.04	0.789
16	8.0	1,600	75	1,525	12.71	0.794
18	9.0	1,800	75	1,725	14.375	0.799
20	10.0	2,000	75	1,925	16.04	0.802

Here $a = 120,900$ $\tau = 200,000$ $f = 75,000$, so that with $i = 20$, $n = \dfrac{10 \times 200,000 \times 75,000}{120,000} \, 1 = 6.04$.

400. It will be observed that, after the first two lamps have been lighted, the speed of the motor is nearly proportional to the number of lamps, and, therefore, the total number of revolutions of a motor armature in a given time, will be an approximate measure of the total quantity of electricity supplied through the meter in coulombs, or in ampere-hours.

In order that the error, introduced into the indications of the meter, by constant friction of the armature, shall be as small as possible, it is important that the constant torque-per-revolution-per-second shall be as great as possible, relatively to the friction, or that $\dfrac{f}{a}$ shall be a small fraction.

401. In practice it would be very difficult to arrange a motor of this kind, having its armature placed directly in the main circuit of the lamps, for the reason that if the brushes were sufficiently fine to permit the friction of the armature to become negligibly small, any accidental short-circuit, occurring between the lamp-leads, would probably destroy the brushes, or the armature, or both. The problem has, however, been successfully met in practice by making the armature in this case the fixed element of the motor, and the field magnet the moving element.

Fig. 228 represents a well-known type of meter, in which the current to be measured passes through the stationary element of the field coils F, F, while the moving element, or armature M, is permanently magnetized by a feeble current passing through a comparatively high resistance, wound on a frame at the back of the instrument and kept in circuit with the mains. The armature M, receives its current through the delicate brushes b, which rest on opposite sides of a small silver commutator c. No iron is employed in either the field or armature of the apparatus. The vertical shaft of the armature M, is geared directly with a dial-recording mechanism similar to that of a gas meter. In order to apply a load torque proportioned to the speed, a disc of copper D, is mounted horizontally upon the vertical armature axis, so as to rotate between the poles of the three permanent magnets P, P, P, as shown. When the disc is at rest there is no retarding torque other than a small mechanical friction due to the brushes resting on the commutator and the weight of the armature in its bearings.

As soon as the disc is set in motion by the rotation of the armature, eddy currents are produced in its substance by the dynamo action of the permanent magnets upon it, and a retarding torque is set up between the disc and these magnets. At all ordinary speeds this torque is proportional to the rate of rotation, thus complying with the requirements of the motor as a meter.

402. The armature of the motor represented in Fig. 227 is

FIG. 228.—WATTMETER.

only capable of acting as a *coulomb meter*, or *ampere-hour meter*, but the apparatus shown in Fig. 228, while acting as an ampere-hour meter on constant potential mains, also operates as a *watt-meter*, in cases where the pressure between the mains is not constant; for, all variations in the pressure will also increase in direct proportion the useful flux Φ, linked with field and armature, and so the speed of the armature will be accelerated and retarded in proportion to the pressure, as well as in proportion to the current strength.

403. No law of retarding torque, other than a torque proportional to the speed, can give a rate of revolution in the armature proportional to the current strength passing through it, when the field flux Φ is constant. If, however, the field magnets be in series with the armature, so that Φ increases with the load, it is possible for an instrument of this character to register fairly accurately, even although the load torque is not proportional to the speed. In such cases, however, the results can only be approximate, since the hysteresis in the magnetic circuit of the field will bring about a complicated relation between load and flux.

404. Another problem which sometimes arises, is to design a motor whose speed shall be proportional to the pressure in

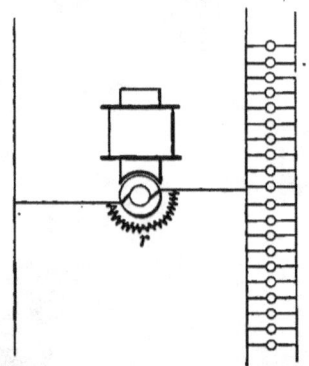

FIG. 229.—MOTOR ARMATURE SHUNTED AND IN CIRCUIT WITH INCANDESCENT LAMPS.

volts at its terminals. This problem presents itself in motor-meters having an armature which, instead of being inserted directly in the lamp circuit, is shunted by a constant small resistance r. A motor-meter of this type is shown in Fig. 229. Here the danger of burning out the armature by an accidental overload is not nearly so great, since the pressure at the armature terminals can never exceed that of the drop in the shunt resistance r. If i, be the total current strength in amperes passing through the lamps, and e, the dynamo power of the armature, in volts-per-revolution-per-second, the current strength passing through the armature will be

$$i_1 = \frac{(i-i_1)r - ne}{R} \qquad \text{amperes,} = \frac{ir \times ne}{R+r}$$

where R is the resistance of the motor armature in ohms, and the driving torque will be $i_1 \tau$ cm.-dynes.

If the frictional torque f, centimetre-dynes, be assumed constant, the speed of the motor will be determined by the relation $i_1 \tau = f$ or

$$\tau \frac{(i r - n e)}{(R + r)} - f,$$

from which $n = \dfrac{i r}{e} - \dfrac{f (R + r)}{e \tau}$ revolutions-per-second.

From this it will be seen that the motor will develop a speed

Lamps.	Current thro' lamps, amperes i.	Current thro' armature, amperes i_1.	Current in shunt amperes $(i-i_1)$.	Drop in shunt, volts.	Drop in armature, volts.	C.E.M.F. of armature, volts.	Speed, revolutions per sec., n.	Revolutions per lamp per second.
1	0.5	0.05	0.45	0.045	0.005	0.040	06.6	0.66
2	1.0	0.05	0.95	0.095	0.005	0.090	1.5	0.75
4	2.0	0.05	1.95	0.195	0.005	0.190	3.16	0.79
6	3.0	0.05	2.95	0.295	0.005	0.290	4.83	0.805
8	4.0	0.05	3.95	0.395	0.005	0.390	6.5	0.813
10	5.0	0.05	4.95	0.495	0.005	0.490	8.16	0.816
20	10.0	0.05	9.95	0.995	0.005	0.990	16.5	0.825

proportional to the main current i, if the frictional torque f, be constant, and sufficiently small to make $\dfrac{f(R+r)}{e \tau}$ small compared with $\dfrac{i r}{e}$. The following case will illustrate this result.

Let $R = 0.1$ ohm, $r = 0.1$ ohm, $f = 50$ cm.-gms., $\tau = 1{,}000$ cm.-gms.-per-ampere, $e = 0.06$ volt per revolution per second. Then $n = 1.667 i - 0.1667$. The preceding table shows the results which follow for various currents up to 10 amperes, either directly from the formula or by independent reasoning.

Such a motor will usually operate at a comparatively high speed at full load, since it depends upon the influence of its C. E. M. F. in reducing the current strength through the armature to that required in order just to balance the resisting torque f.

405. If, however, a load torque be imposed on the armature, proportional to the speed, represented by $\tau_1 = a n$, then our relation becomes

$$i_1 = r f + a n$$

$$\frac{ir-ne}{R+r}\tau = f + an, \text{ from which } n = \frac{ir\tau - f(R+r)}{e\tau + a(R+r)}$$

revolutions-per-second.

If, for example, in the last case, the motor develops a retarding torque of 60 cm.-gms. per-revolution-per-second (a = 60 cm.-gms. or 60,000 cm.-dynes approximately), we obtain either from the formula, or by direct analysis, the following results:

	Current.				Torque.				Drop in Motor Armature.			Speed.	
No. of lamps.	Thro' lamp amperes.	Thro' motor amperes.	Thro' shunt amperes.	Drop of pressure in shunt, volts.	Friction, cm.-gms.	Speed, cm.-gms.	Total, cm.-gms.	Current thro' motor amperes.	Due to resistance, volts.	C. E. M. F., volts.	Total volts.	Rv. p. sec.	Rv. p. sec. per lamp.
1	0.5	0.083	0.417	0.0417	50	33	83	0.083	0.0083	0.033	0.0413	0.55	0.55
2	1	0.125	0.875	0.0875	50	75	125	0.125	0.0125	0.075	0.0875	1.25	0.625
4	2	0.208	1.792	0.1792	50	158	208	0.208	0.0208	0.158	0.179	2.633	0.658
6	3	0.292	2.708	0.2708	50	242	292	0.292	0.0292	0.242	0.271	4.03	0.667
8	4	0.375	3.625	0.3625	50	325	375	0.375	0.0375	0.325	0.3625	5.42	0.678
10	5	0.459	4.541	0.4541	50	409	459	0.459	0.0459	0.409	0.454	6.82	0.682
20	10	0.876	9.124	0.9124	50	826	876	0.876	0.0876	0.826	0.913	13.97	0.689

406. It is, therefore, evident that a motor armature, with constant field excitation, can develop a speed closely proportional to the pressure at its terminals, and, therefore, serve as a motor-meter, if the retarding torque be small and constant, or, if it be partly small and constant, and partly proportional to the speed.

407. One of the most important recent applications of motors is their distributed application to machine tools in large factories. Instead of employing long lines of countershafting, which must necessarily be constantly driven during working hours, a separate electric motor is applied directly to each machine, so that each machine is started and stopped according to its own requirements. Moreover, the range of regulation of speed, which is obtainable from a common countershafting, is necessarily more limited in degree than that which can be effected by the use of independent motors.

408. By the use of *individual electric motors*, not only is each tool capable of operation at its best speeds, and under complete control, but also the friction of long lines of countershafting is eliminated. The economy is greatest where the nature of the work in the machine shop is such that the average power supplied to the tools is much less than the maximum power, or the ratio of average to maximum power; *i. e.*, the *load factor* is small, since the motors, when completely disconnected from a circuit, take no power, whereas, the countershafting consumes, practically, the same amount of power friction, whether the tools be active or idle.

CHAPTER XXX.

MOTOR·DYNAMOS.

409. The consideration of dynamos and motors naturally leads to that of a third class of apparatus, which partakes of the nature of each; namely, *motor-dynamos*, or, as they are sometimes called, *dyna-motors*. It is evident that if a motor be rigidly connected to a dynamo, either by a belt or by a coupling, that we obtain a means whereby electric power can be transformed through the intermediary of mechanical power. Thus, the motor may be operated from a high-tension circuit, while the dynamo operates a low-tension circuit, or *vice versa ;* but, neglecting losses taking place in the two machines, the amount of electric energy absorbed and delivered in the respective circuits will be the same, the combination being utilized for the purpose of transforming the pressure and current strength. For this reason a motor-dynamo is commonly called a *rotary transformer*, in order to distinguish it from an ordinary alternating-current transformer, which always remains at rest.

410. Instead of rigidly connecting together two separate machines; *i. e.*, two armatures in two separate fields, the plan has been adopted of placing the two armatures in a field common to both; as, for example, by placing them in a common field of double length. Or, a still closer union can be effected by winding both the armature and motor coils on a common armature core, care being taken to insulate the two sets of windings from each other. Under these circumstances, since the intake of the motor winding is practically equal to the output of the dynamo winding, the space occupied by each winding will be practically the same, so that where both are associated on a common core, half the winding space is appropriated for each. The result will be that if the motor winding or dynamo winding be such as would appertain to, say, a 10-KW capacity, the armature in which the two are associated will be a machine having, approximately, the size and weight corresponding to a 20-KW capacity. There is, however, an econ-

FIG. 230.—STEP-UP MOTOR DYNAMO.

omy in constructing one machine of double capacity, instead of two machines of single capacity, both in first cost and in efficiency.

411. Rotary transformers, like all transformers, may be either of the *step-up* or *step-down* type. Fig. 230 represents a step-up rotary transformer of 1.5-KW capacity, transforming from 120 volts and 12.5 amperes, to 5,000 volts and 0.3 ampere. The motor winding of the armature is connected with the commutator on the left, while the generator winding of the armature is connected with the commutator on the right. The magnet coils are excited from the low-tension mains. The two armature windings, in such cases, may be either placed one below the other, or they may be interspersed. The left hand brushes receive the 120-volt pressure, and the right hand brushes deliver the 5,000-volt-pressure. The function of such a machine is to test high-tension insulation under practical conditions of pressure.

412. Fig. 231 represents a step-down rotary transformer for transforming from 500 to 120 volts. In this case the smaller

FIG. 231.—STEP-DOWN DYNAMO.

brushes are connected to the 500-volt mains, as is also the field winding, and the lower pressure is delivered at the heavy brushes.

413. It is important to observe that in a motor dynamo of the preceding types there is no appreciable armature reaction. The reason for this is as follows: The M. M. F. of the motor armature winding is, as we know, opposite in direction of that of the generator winding; and, since these M. M. Fs. are nearly equal, and are produced on the same core, they will nearly neutralize each other. Consequently, the brushes of such a machine never require to be shifted during variations of load, and the commutators are characterized by quiet and sparkless operation.

414. Under ordinary circumstances it is necessary to excite the field magnet of a motor dynamo from the primary circuit, since, otherwise, the motor side could not be operated. It is often possible, however, to place a series winding on the motor side, and a shunt winding on the secondary or dynamo side. Thus, if it be required to transform from 1,000 to 50 volts, a shunt-field winding for 1,000 volts would be more expensive than one for 50 volts. In such a case it becomes possible to excite the fields by a few turns of series winding, carefully insulated, in the primary circuit, in order to start the machine from rest, and to supply the balance of the field excitation by a shunt winding on the secondary side, which commences to be actuated as soon as the motor starts.

415. It will be evident that any variation in the strength of the field magnets, whether these be shunt- or series-wound, will not vary the ratio of transformation; for, although by varying the field excitation the motor can be made to change speed, yet this speed will not produce any appreciable effect upon the generated E. M. F., since the field is proportionally weakened. In other words, the C. E. M. F. in the motor being always equal to the E. M. F. at the brushes, after deducting the drop in the armature, the generated E. M. F., which is always some fixed fraction of the motor C. E. M. F., must be constant within the same limits. If the number of turns in the motor winding, counted once all round the armature, be w_m, and the number of turns in the generator winding, counted in the same manner, be w_g, then the ratio $\frac{w_g}{w_m}$ is called the *ratio of transformation*. If, then, the primary E. M. F. be E_1 volts, the primary current I_1 amperes, and the resistance in the primary winding r_1 ohms, while the corresponding quantities in the secondary circuit are E_2, I_2, and r_2, respectively, the C. E. M. F. in the primary winding will be $n_1 e_1 = E_1 - I_1 r_1$, where n_1 is the speed of revolution in turns-per-second, and e_1, the dynamo power, or $\Phi w_m \times 10^{-8}$. The generated secondary E. M. F. will be $n \, \Phi w_g \times 10^{-8}$ volts $= (E_1 - I_1 \, r_1) \dfrac{w_g}{w_m}$.

The pressure at the secondary terminals will be further reduced by the drop in the secondary winding; or

$$E_2 = (E_1 - I_1 \, r_1) \dfrac{w_g}{w_m} - I_2 \, r_2.$$

If the weight of copper in the two windings is equal, $I_2 \, r_2$, will practically be equal to $\dfrac{I_1 \, r_1 \, w_g}{w_m}$, so that

$$E_2 = E_1 \dfrac{w_g}{w_m} - 2 I_2 \, r_2.$$

The machine, therefore, acts as though it were a dynamo of E. M. F. $\dfrac{w_g}{w_m} E_1$, with an internal resistance of $2r_2$, or twice that of the secondary winding.

416. In all motor dynamos, having a field magnet common to both armatures, the ratio of transformation, neglecting armature drop, is constant, no matter how the field excitation is varied. Motor-generators are often employed for raising or lowering the pressure of continuous-current circuits. Thus electroplating E. M. Fs. of, say 6 volts, are obtainable in this manner from circuits of 110, 220 or 500 volts pressure. Similarly, pressure of 150 volts are obtainable from a few storage batteries by such apparatus.

417. In central stations for low-pressure distribution, say at 220 volts, by a three-wire system, some of the feeders have to be maintained at a higher pressure than others, in order that all the *feeding points*, or points of connection between feeders and the mains, should have the same pressure. This is accomplished either by employing separate dynamos, operated at slightly different pressures, or by introducing at the central station motor-dynamos having their dynamos in circuit with the feeders. Such motor-dynamos are frequently called *boosters*. The motor-dynamo for this purpose requires that means should be provided for regulating the E. M. F. which is to be added to the feeder circuit. This can only be done by employing separate field magnets for the motor and generator armatures.

MOTOR DYNAMOS. 323

Fig. 232 represents a practical form of booster employed in a three-wire central station. The middle machine is a motor operated at central-station pressure of, perhaps, 250 volts; the others are generators, having their armatures coupled to the same shaft as that of the motor armature. One dynamo is

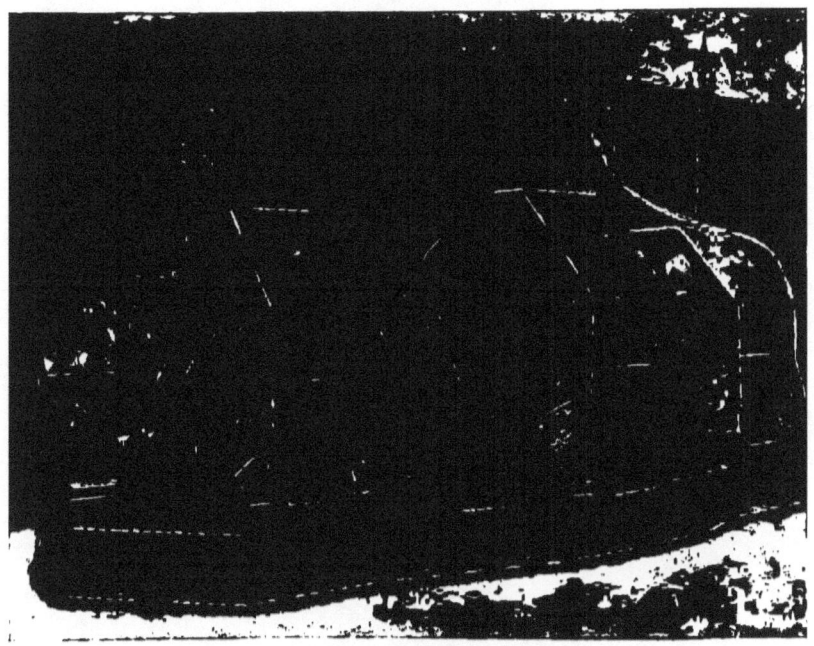

FIG. 232.—BOOSTER IN THREE-WIRE CENTRAL STATION.

connected in circuit with the positive conductor of the feeder whose pressure is to be raised, and the other is connected in the circuit of the negative conductor. Since these feeders carry heavy currents and require to be of very low resistance, the necessity for the massive copper brushes and connections of the dynamos will be evident. The amount of E. M. F. which will be generated in these armatures will be determined by the excitation of their field magnets.

THE END.

INDEX.

Active Conductor, Magnetic Flux of, 37
Aero-Ferric Magnetic Circuits, 68–73
Air-Gap, Magnetic, 57
Air-Path, Alternative Magnetic, 42
— Aligned M. M. F., 56
Alternating-Current Dynamos, 17
Alternative Magnetic Air-Path, 52
Alternators, 17
— Multiphase, 25
— Uniphase, 26
Ampere, Definition of, 49
Ampere-Hour Meter, 313
Ampere-Turn, Definition of, 40
Anomalous Magnet, 47
Arc-Light Dynamos, 26
Armature, Back Magnetization of, 186
— Cores, Cross-Sections of, 126
—, Core Discs for, 152
— Core, Lamination of, 105
—, Cylinder or Drum, 23
— Disc, 23
—, Double Winding of, 190
—, Gramme-Ring, 23
—, I² R, Loss in, 200
—, Iron-Clad, Definition of, 24
—, Journal Bearings, 159–163
— of Machine, 9
—, Neutral Line of, 184
—, Pole, 110–116
—, Radial, 110
— Reaction and Sparking at Commutators, 179–198
— Ring, 23
—, Smooth-Core, 23, 152
— — Definition of, 24
— Toothed-Core, 152
— —, Definition of, 24
— —, 23
— Turns, Effect of, on E. M. F., 3
— Winding, Closed-Coil, 110
— —, Disc, 230
— —, Dissymmetry of, 125
— —, Inter-Connected, 145
— — Space, 275
— Wire, Effective Length of, 246

Armatures, Closed-Coil, 217
—, Gramme-Ring, 117–127
—, Lap Winding for, 155
—, Open-Coil, 217
—, Wave-Winding for, 155
Attractions and Repulsions, Laws of Magnetic, 33
Automatic Regulation of Dynamos, 218
Average Efficiency of Motor, 279

Back Magnetization of Armature, 186
Balancing Coil of Armature, 194
Bar, Equalizing, 224
Bars, Bus, 224
—, Omnibus, 224
Bearings, Self-Oiling, 161
Belt-Driven Dynamos, 18, 135
Bipolar Dynamo, 16
Boosters, 322, 323
Box, Field-Regulating, for Dynamo, 14
Brush, Dynamo, 124
Brushes, Forward Lead of, 217
—, Lead of, 185
— of Dynamo, 9
— of Motor, Lag of, 303
Bus Bars, 224

Calculation of Gramme-Ring Dynamo Windings, 128–134
Capability, Electric, of Dynamo, 126
—, Electric, of Dynamo-Electric Machine, 4
Car Motor, 277
Characteristic Curve of Dynamos, 210
— External, of Series-Wound Dynamo, 210
— Internal, of Series-Wound Dynamo, 210
— of Shunt-Wound Dynamo, 212
Circuit, Magnetic, 48
— Return, for Track Feeders, 226
Circuits, Ferric-Magnetic, 55–67
—, Magnetic, Non-Ferric, 48–54

INDEX.

Circuit, Transmission, Definition of, 1
Circular Distribution of Magnetic Flux Around Conductor, 37
—, Magnetic Flux, Assumed Direction of, 39
Closed Circular Solenoid, 50
— Coil Armature Winding, 110
— — Armatures, 217
Coefficient, Hysteretic, 174
Coil, Balancing, of Armature, 194
—, Inductance, 301
—, — of, 181
—, Starting, 301
Combinations of Dynamos in Series or in Parallel, 220-227
Commercial Efficiency of Dynamo, 5
— — of Dynamos, Circumstances Affecting, 7
— — of Motor, 268
Commutation, Definition of, 180
—, Diameter of, 180
—, Quiet, Circumstances Favoring, 187
—, Sparkless, Circumstances Favoring, 186
Commutator, Circumstances Favoring Sparking at, 186
—, Forms of, 123
— of Dynamo, 9
Commutatorless, Continuous-Current Dynamo, Disc Type of, 236
— — Dynamos, 234
— — Generators, 234-240
Commutators, Sparking at, 179-198
Compound Magnets, 105
— Winding of Dynamos, 208
Compound-Wound Dynamos, 14
— —, Uses for, 209
Conductor, Active, Magnetic Flux of, 37
Consequent Poles of Dynamo, 22
Constant-Current Dynamos, 10
Constant-Potential Dynamos, 10
Constants, Reluctivity, Table of, 65
Continuous-Current Commutatorless Dynamos, 28, 234
— — —, Cylinder Type of, 236
— — Dynamo, 20
— — Generators, 234-240
— — Generator, Limitations to Output of, 203
Convention as to Direction of Circular Magnetic Flux, 39
Converging Magnetic Flux, 35
Core Discs for Armatures, 152

Core, Effect of Lamination on Eddy Currents, 166
Coulomb Meter, 313
Counter Electro-Dynamic Force, 256
Cross Magnetization, 183
Currents, Eddy, 164-171
—, Eddy, Definition of, 164
—, —, Effect of Lamination of Core on, 166
—, —, Origin of, 165
Curves, Characteristic of Dynamos, 210
— of Reluctivity in Relation to Flux Density, 66
Cutting Process vs. Enclosing of Magnetic Flux, 82
Cycles of Magnetization, 174
Cylinder or Drum Armature, 23
— Type of Commutatorless Continuous-Current Dynamos, 236

Decipolar Dynamos, 17
Density, Flux, 34
—, Prime Flux, 54
Devices, Receptive, Definition of, 1
Diameter of Commutation, 180
Diffusion, Magnetic, 52, 53
Diphase Dynamo, 27
Direct-Driven Dynamos, 135
Disc Armature, 23
— Armature Winding, 230
— Armatures and Single Field-coil Machines, 228-233
—, Faraday's, 234
— Type of Commutatorless Continuous-Current Dynamos, 236
Dissymmetry, Magnetic, 124
— of Armature Winding, 125
Distribution of Magnetic Field, 41-47
— of Magnetic Flux, 31
— of Magnetic Flux of Conductor, 37
Diverging Magnetic Flux, 35
Double Circuit, Bipolar Dynamo, 16
Double Winding of Armature, 190
Drum Armatures, 152
— or Cylinder Armatures, 23
Dynamo Armatures, Electro-Dynamic Induction in, 90-102
—, Bipolar, 16
—, Brush, 124
—, Brushes of, 9
—, Commercial Efficiency of, 5
—, Commutator, 9
—, Consequent Poles of, 22

INDEX. 327

Dynamo, Continuous-Current, 20
—, Diphase, 27
—, Double-Circuit, Bipolar, 16
—, Electric Capability of, 126
—, — Efficiency of, 5
Dynamo-Electric Generator, 2
— — Machine, Electric Capability of, 5
Dynamo Field-Regulating Box, 14
—, Intake, 5
—, Load of, 15
—, Magneto-Electric, 11
—, Output of, 5
—, Plating, 26
Dynamo-Power of Motor, 266
Dynamo Relation between Output and Resistance, 6
—, Self-Excited, 12
—, —, Compound-Wound, 13
—, Separately Excited, 12
—, Single-Circuit, Bipolar, 16
—, Telegraphic, 26
Dynamos, Alternating-Current, 17
—, Arc-Light, 26
—, Automatic Regulation of, 218
—, Belt-Driven, 18
—, Characteristic Curves of, 210
—, Circumstances Influencing Electric and Commercial Efficiency of, 7
—, Combination of, in Series or Parallel, 220–227
—, Commutatorless Continuous-Current, 28, 234
—, Compound-Wound, 14
—, —, Uses for, 209
—, Constant-Current, 10
—, Constant-Potential, 10
—, Decipolar, 17
—, Direct-Driven, 135
—, Heating of, 199–205
—, Incandescent Light, 26
—, Inductor, 25
—, Multipolar, 16
—, Multipolar, Gramme-Ring, 135–151
—, Octopolar, 17
—, Over-Compounded, 209
—, Quadripolar, 17
—, Regulation of, 206–219
—, Self-Excited, Series-Wound, 13
—, Series-Wound, Uses for, 209
—, Sextipolar, 17
—, Shunt-Wound, Uses for, 209
—, Simple Magnetic Circuits, 22
—, Single-Field-Coil, Multipolar, 28
—, Single-Phase, 27
—, Three-Phase, 27

Dynamos, Triphase, 27
—, Two-Phase, 27
—, Unipolar, 28
Dynamotors, 317
Dyne, Definition of, 69

E. M. F., Effect of Number of Armature Turns on, 3
—, Effect of Speed of Revolution on, 3
—, Induced by Magneto Generators, 103–109
—, Induced in Loop, Rule for Direction of, 94
—, of Electro-Dynamic Induction, Value of, 75–82
—, of Self-Induction, 181
—, of Self-Induction, Circumstances Affecting Value of, 182
—, Produced by Cutting Earth's Flux, 90
Earth's Flux, E. M. F. Produced by Cutting, 90
Eddy Currents, 164–171
— —, Definition of, 164
— —, Effect of Lamination of Core on, 166
— —, Formation of, in Pole-pieces, 169
— —, Origin of, 165
Edges, Leading, of Pole-pieces, 184
Efficiency, Average, of Motor, 270
—, Full Load of Motor, 270
— of Motors, 268–279
Electric Capability of Dynamo, 126
— — of Dynamo-Electric Machine, 5
— Efficiency of Dynamos, Circumstances Affecting, 7
— Flux, Unit of, 49
Electro-Dynamic Force, 241–249
— Induction, 75–82
— — in Dynamo Armature, 90–102
— —, Laws of, 74–89
— Machinery, 1
— Machinery, Classification of, 1
Enameled Rheostats, 216
Entrefer, 105
Equalizing Bar, 224
Ether, Assumed Properties of, 29
Ether Path of Reluctivity, 60
External Characteristic of Series-Wound Dynamo, 210

Factor, Leakage, 132
—, Load, 317
Faraday's Disc, 234
Feeders for Return Track, 226

INDEX.

Feedin Points, 322
Ferric Magnetic Circuits, 55–67
— Path of Metallic Reluctivity, 60
Field Magnet of Machine, 9
—, Magnetic, 32
— Magnets, I^2R Losses in, 199
— Poles, Eddy-Current Losses in, 200
— Regulating Box for Dynamo, 14
— Rheostats, 215
Fleming's Hand Rule for Dynamos, 74
— — — — Motors, 243
Flux, Circular Magnetic, Convention as to Direction of, 39
—, Converging Magnetic, 35
— Density, 34
—, Diverging Magnetic, 35
—, Magnetic, Unit of, 49
— Density, Prime, 54
—, Prime, 56
—, Magnetic, 29
—, —, Distribution of, 31
—, —, Irregular, 35
—, —, Variations of, 33
— Paths, Magnetic, 2
Following Edges of Pole-Pieces, 184
Force, M. M., Induced, 56
—, Electro-Dynamic, 241–249
—, Lines of Magnetic, 34
—, Magnetic, Tubes, 35
—, Magnetizing, 53
—, Magnetomotive, 31
—, —, Prime, 56
Forces, Electromotive, Methods for Increasing, 3
French Measures, Table of, 8
Friction Losses in Bearings and Brushes, 201
—, Magnetic, 174
Full-Load Efficiency of Motor, 270

Gap, Magnetic Air, 57
Gauss, Definition of, 35
Generator Armature, Limiting Temperature of, 203
—, Dynamo-Electric, 2
Generators, Commutatorless Continuous-Current, 234–240
—, Definition of, 1
Gilbert, Definition of, 40
Gramme-Ring Armature, 23
— Armatures, 117–127
— Dynamos, Multipolar, 135–151

Hand Rule, Fleming's, for Dynamos, 74

Heating of Dynamos, 199–205
Hysteretic Activity, Table of, 175
— Losses in Armature and Field Poles, 200
— Loss, 174
— Coefficient, 174
Hysteresis, Magnetic, 172–178
—, —, Definition of, 172

Incandescent Light Dynamos, 26
Individual Electric Motors, 317
Idle Wire on Armature, 100
Inductance Coil, 301
— of Coil, 181
Induction, Electro-Dynamic, 75–82
—, —, Laws of, 74–89
— in Dynamo Armature, 90–102
—, Self, E. M. F. of, 181
Inductor Dynamos, 25
Intake of Dynamo, Definition of, 5
Inter-Connected Armature Winding, 145
Internal Characteristics of Series-Wound Dynamo, 210
Iron-Clad Armature, 24
Irregular Magnetic Flux, 35

Joint Reluctivity, 60
Journal Bearings for Armatures, 159–163

Lag of Motor Brushes, 303
Lamination of Armature Core, 105
Lamp, Pilot, Definition of, 12
Lap Winding for Armatures, 155
Laws of Electro-Dynamic Induction, 74–89
— — Magnetic Attractions and Repulsions, 33
Lead, Forward, of Dynamo Brushes, 217
— of Brushes, 195
Leading Edges of Pole-pieces, 184
— Pole of Motor, 303
Leakage Factor, 132
—, Magnetic, 52, 53
Length, Effective, of Armature Wire, 246
Limitation to Output of Continuous-Current Generator, 203
Limiting Temperature of Generator Armature, 203
Line, Neutral, of Armature, 194
Lines of Magnetic Force, 34
—, Stream, 30
Load Factor, 317
— of Dynamo, 15
Locomotors, 273

INDEX.

Loss by Eddy Currents in Armature and Field Poles, 200
—, Hysteretic, 174
—, —, in Armature and Field Poles, 200
Losses, $I^2 R$, in Field Magnets, 199
— in Armature, $I^2 R$, 200
— Produced by Air-Churning, 201
— — — Friction in Bearings and Brushes, 201

M. M. F., Aligned, 56
—, Induced, 56
—, Methods of Producing, 38
—, Prime, 56
—, Structural, 56
—, Unit of, 40
Machine, Armature of, 9
— Circumstances Influencing Electric Efficiency of Dynamos, 7
—, Field Magnet of, 9
—, Magnetic Flux Produced by, 9
Machinery, Electro-Dynamic, 1
—, —, Classification of, 1
Machines, Disc Armature and Single Field-Coil, 228–233
Magnet, Anomalous, 47
—, Mechanical Analogue of, 30
—, North-Seeking Pole of, 29
Magnets, Compound, 105
—, Molecular, 56
Magnetic Air-Gap, 57
— Air Path, Alternative, 52
— Attractions and Repulsions, Laws of, 33
— Circuit, 48
— Circuit, Application of Ohm's Law to, 49
— Circuits, Aero-Ferric, 68–73
— Diffusion, 52, 53
— Field, 32
— Dissymmetry, 124
— Field, Distribution of, 41–47
— —, Method of Mapping, 32
— —, Negatives of, 32
— —, Photographic Positives of, 32
— Flux, 29
— —, Converging, 35
— —, Cutting Process, Enclosing, 82
— — Density, 34
— —, Diverging, 35
— —, Effect of, on C. E. M. F., 58
— —, Irregular, 35
— — of Dynamo, 9
— —, Uniform, 35
— —, Unit of, 49
— —, Unit of Intensity of, 35

Magnetic Flux, Variations of, 33
— Force, Tubes of, 35
— Friction, 174
— Hysteresis, 172–178
— —, Definition of, 172
— Intensity, 34
— Leakage, 52, 53
— Permeability, 55
— —, Definition of, 3
— Potential, Fall of, 53
— Reluctance, 48
Magnetism, Definition of, 29
—, Molecular, 56
—, Residual, 55, 173
—, Streaming-Ether Theory of, 29
Magnetization, Back, of Armature, 186
—, Cross, 183
—, Cycles of, 174
Magnetizing Force, 53
— — in Relation to Reluctivity, 59
Magneto-Electric Dynamo, 11
Magneto Generators, E. M. F. Induced by, 103–109
Magnetomotive Force, 31
Mapping of Magnetic Field, 32
Mechanical Analogue of Magnet, 30
Meter Motors, 309–317
Methods for Suppressing Sparking, 189
Molecular Magnetism, 56
— Magnets, 56
Motor, Average Efficiency of, 270
—, Commercial Efficiency of, 268
—, Dynamo-Power of, 264
— Dynamos, 318–323
—, Definition of, 318
—, Full-Load Efficiency of, 270
—, Leading Pole of, 303
— Torque, 251–267
—, Trailing Pole of, 303
Motors, Efficiency of, 268–279
—, Fleming's Hand Rule for, 243
— for Street Car, 277
—, Individual Electric, 217
—, Regulation of, 280–296
—, Slow Speed, 271
—, Starting and Reversing of, 291–308
—, Stationary, 273
—, Traveling, 273
Multiphase Alternators, 26
Multipolar Dynamos, 16
— —, Single-Field-Coil, 28
— Gramme-Ring Dynamos, 135–151

Negatives of Magnetic Fields, 32
Neutral Line of Armature, 184
— Wire of Three-Wire System, 221
Non-Ferric Magnetic Circuits, 48–54
North-Seeking Pole of Magnet, 29

Octopolar Dynamos, 17
Oersted, Definition of, 49
Ohm, Definition of, 49
Ohm's Law, 49
— — Applied to Magnetic Circuit, 49
Oilers, Sight-Feeding, 160
Omnibus Bars, 224
Open-Coil Armatures, 217
Over-Compounded Dynamos, 209
Output and Dimensions of Dynamos, Relation Between, 136
— of Dynamo, Definition of, 5
— —, Relation Between and Resistance, 6

Permeability, Magnetic, 55
—, —, Definition of, 3
Photographic Positives of Magnetic Fields, 32
Pilot Lamp, Definition of, 12
Plating Dynamo, 26
Points, Feeding, 322
Pole Armature, 25
— Armatures, 110–116
—, Leading, of Motor, 303
—, North-Seeking of Magnet, 29
—, South-Seeking, 29
—, Trailing, of Motor, 303
Pole-Pieces, Following Edges of, 184
—, Formation of Eddy Currents in, 169
—, Leading Edges of, 184
Poles, Consequent, of Dynamo, 22
Potential, Magnetic, Fall of, 53
Prime Flux, 56
— Flux Density, 54
— M. M. F., 56
Properties, Assumed, of Ether, 29

Quadripolar Dynamos, 17
Quiet Commutation, Circumstances Favoring, 187

Radial Armature, 110
Ratio of Transformation, 321
Receptive Devices, Definition of, 1
Regulation of Dynamos, 206–219
— of Motors, 280–296
Reluctance, 48
—, Magnetic, 48

Reluctance, Unit of, 49
Reluctivity, 48
—, Constants, Table of, 65
— Curves in Relation to Flux Density, 66
—, Ether Path of, 60
— in Relation to Magnetizing Force, 59
—, Joint, 60
—, Metallic, Ferric Path of, 60
Residual Magnetism, 55, 173
Resistivity, 48
Return Track Feeders, 226
Reversing and Starting of Motors, 291–308
Rheostats, Enameled, 216
—, Field, 215
—, Starting, 298
Ring Armature, 23
Ring Armatures, Gramme, 117–127
Rotary Transformers, 318
Rule, Fleming Hand, for Motors, 243
— for Direction of E. M. F. Induced in Loop, 94

Self-Excited Compound-Wound Dynamo, 13
— Dynamo, 12
— Series-Wound Dynamos, 13
Self-Induction, E. M. F., of, 181
— E. M. F., of, Circumstances Affecting Value of, 182
Self-Oiling Bearings, 161
Separately-Excited Dynamo, 12
Series or Parallel Combinations of Dynamos, 220–227.
— Winding of Dynamos, 206
Series-Wound Dynamo, External Characteristic of, 210
— Dynamo, Internal Characteristic of, 210
Sextipolar Dynamo, 17
Shunt Winding of Dynamos, 207
Shunt-Wound Dynamo, Characteristic of, 212
— Dynamos, Uses for, 209
Sight-Feeding Oilers, 160
Simple Magnetic Circuit Dynamos, 22
Single-Circuit Bipolar Dynamo, 16
Single Field-Coil Multipolar Dynamos, 28
Single-Phase Dynamos, 27
Slow Speed Motor, 271
Smooth-Core Armature, 23
— Armatures, 152

Smooth-core Armature, Definition of, 24
Solenoid, Closed Circular, 50
Sources, Electromotive, 2
South-Seeking Pole, 29
Space for Armature Winding, 275
Sparking and Armature Reaction, 179–198
— at Commutator, Circumstances Favoring, 186
—, Definition of, 180
—, Methods for Suppressing, 189
Sparkless Commutation, Circumstances Favoring, 186
Specific Resistance, 48
Speed of Revolution, Effect of, on E. M. F., 3
Starting and Reversing of Motors, 291–308
— Coil, 301
— Rheostats, 298
Stationary Motors, 273
Step-Down Transformers, 319
Step-Up Transformers, 319
Stream Lines, 30
Streaming-Ether Theory of Magnetism, 29
Structural M. M. F., 56
System, Three-Wire, 221

Table of French Measures, 8
— of Hysteretic Activity, 175
— of Reluctivity Constants, 65
Telegraphic Dynamo, 26
Thermal Losses, 204
Three-Phase Dynamos, 27
Three Phasers, 27
Three-Wire System, 221
— —, Neutral Wire of, 221
Toothed-Core Armature, 23
— —, Definition of, 24
— Armatures, 152
Torque, Definition of, 251
—, Motor, 251–267
Transformation, Ratio of, 321
Transformers, Rotary, 318
—, Step-Down, 319
—, Step-Up, 319
Transmission Circuits, Definition of, 1

Travelling Motors, 273
Triphase Dynamos, 27
Triphasers, 27
Tubes of Magnetic Force, 35
Turns, Armature, Effect of, on E. M. F., 3
Two-Phase Dynamos, 27
Two Phasers, 27

Uniform Magnetic Flux, 35
Uniphase Alternators, 26
Unipolar Dynamos, 28, 234
Unit of Electric Flux, 49
— — Force, in C. G. S. System, 68
— — M. M. F., 40
— — Magnetic Flux, 49
— — — Intensity, 35
— — Reluctance, 49

Variations of Magnetic Flux, 33
Volt, Definition of, 49
Voltaic Analogue of Aero-Ferric Circuit, 69
— — — Simple Ferric Circuit, 69
— Circuit, Magnetic Analogue of, 53

Wattmeter, 313
Wave Winding for Armatures, 155
Weber, Definition of, 49
Winding, Closed-Coil Armature, 110
—, Compound, of Dynamos, 208
—, Disc Armature, 230
— for Armature, Inter-Connected, 145
— — Armatures, Lap, 155
— — Armature, Wave, 155
— of Gramme-Ring Dynamo, Calculations of, 128–134
—, Shunt, of Dynamos, 207
—, Space, for Armature, 275
Wire, Armature, Effective Length of, 246
—, Idle, on Armature, 100
—, Neutral, of Three-wire System, 221

www.ingramcontent.com/pod-product-compliance
Lightning Source LLC
Chambersburg PA
CBHW030002240426
43672CB00007B/799